Introduction and Guide to the Marine Bluegreen Algae

INTRODUCTION AND GUIDE TO THE MARINE BLUEGREEN ALGAE

HAROLD J. HUMM
University of South Florida

SUSANNE R. WICKS
Southern Illinois University

A Wiley-Interscience Publication
JOHN WILEY AND SONS
New York • Chichester • Brisbane • Toronto

Library of Congress Cataloging in Publication Data:

Humm, Harold Judson.
    Introduction and guide to the marine bluegreen algae.

    "A Wiley-Interscience publication."
    Bibliography: p.
    Includes indexes.
    1. Cyanophyta.   2. Marine algae.   3. Cyanophyta—
Identification.   4. Marine algae—Identification.
I. Wicks, Susanne R., joint author.   II. Title.

QK569.C96H85   589'.46   79-24488
ISBN 0-471-05217-5

Printed in the United States of America

10 9 8 7 6 5 4 3 2 1

PREFACE

This book presents a general introduction to the nature of the bluegreen algae and a guide to the identification of those species known to grow in the oceans of the world. Identifications can be made following the newer taxonomy of Dr. Francis Drouet or in accordance with the older taxonomy.

There has been a need for such a guide for a long time, a need that has increased since the publication of Drouet's series of monographs in which he has drastically revised the taxonomy of the Myxophyceae. Although identification, *sensu* Drouet, of the bluegreens found in the sea can be done by using these monographs, it requires the availability of all four of them (the fifth and last is in press), and they were not written specifically for routine identification by students.

Many biologists interested in marine bluegreen algae prefer not to follow Drouet's taxonomy for reasons scientific or otherwise. This guide will allow their identification by the older taxonomy as well. Even those who recognize the value of the Drouet taxonomic system are often interested in the earlier nomenclature. In many instances the older names are of special value in designating ecophenes or certain environmental variants.

Keys, descriptions, and illustrations are provided as well as habitat and distributional notes.

HAROLD J. HUMM
SUSANNE R. WICKS

St. Petersburg, Florida
Lebanon, Illinois
April 1980

CONTENTS

Introduction   1

Class Myxophyceae, the Bluegreen Algae   3
Classification   4
Taxonomy   5
Growth Form and Morphology   6
Cytology   8
   Nucleoplasm and Chromoplasm   8
   DNA and RNA   8
   Genome Replication   10
   Photosynthetic Pigments   10
   Carotinoids   11
Photosynthesis   12
Food Sources   13
Mutation   13
Genetic Recombination   14
Nitrogen Metabolism   15
Storage Products   15
   Myxophycean Starch   16
   Polyphosphate Granules   16
   Cyanophycin Granules   16
   Lipids   17
Nitrogen Fixation   17
   The Heterocyst Species   19
   Nonheterocyst Species   19
The Cell Wall   21
   Lysis of Bacteria   22
The Sheath   22
   Rate of Production   23
   Pigmentation of the Sheath   23

Chemical Nature and Physical Properties 23
Function 24
Motility 24
Reproduction and Dissemination 25
Endospores and Hormogonia 25
Akinetes 26
Heterocysts 27
Development 27
Plasmodesmata 27
Location 29
Wall Resistance 29
Function 29
Gas Vacuoles 29
Cyanophages 33
Distribution in the Sea 34
Intertidal Zone 34
Below Low Tide 35
Below the Euphotic Zone 35
In the Plankton 36
Within Limestone 36
Temperature and Salinity 37
Toxic Bluegreens 38
Collection and Preservation 39
Microscopic Examination 40

CYANOPHYTA 43

Key to the Genera of the Cyanophyta 44
Order Coccogonales 45
Key to the Marine Families of the Order Coccogonales 47
Family Chroococcaceae 47
Key to the Genera of the Chroococcaceae 48
Coccochloris 48
Johannesbaptistia 51
Agmenellum 52
Gomphosphaeria 53
Anacystis 55

Family Chamaesiphonaceae 58
   Entophysalis 60
Order Hormogonales 62
Key to the Families of the Hormogonales 64
Family Oscillatoriaceae 64
   Spirulina 65
   Oscillatoria 66
   Schizothrix 70
   Arthrospira 74
   Porphyrosiphon 76
   Microcoleus 80
Family Nostocaceae 83
Key to the Marine Genera of the Nostocaceae 84
   Calothrix 84
   Scytonema 85
   Anabaina 88
   Nostoc 89
Family Stigonemataceae 90
   Key to the Genera and Species 91
   Mastigocoleus 91
   Brachytrichia 92

APPENDIX   97

Table 1. List of Species, *sensu* **Drouet, and the Synonyms of**
   Each that are Treated in the Appendix 98
Key to the Older Genera of the Cyanophyta 102
   Chroococcus 105
   Merismopedia 106
   Aphanothece 109
   Gloeocapsa 110
   Aphanocapsa 110
   Oncobyrsa 111
   Dermocarpa 111
   Pleurocapsa 114
   Entophysalis 116
   Hyella 118
   Xenococcus 118

Spirulina   119
Trichodesmium   123
Oscillatoria   124
Lyngbya   129
Phormidium   134
Symploca   138
Plectonema   140
Hydrocoleum   142
Microcoleus   144
Schizothrix   145
Amphithrix   146
Calothrix   147
Richelia   153
Fremyella   153
Dichothrix   155
Rivularia   157
Scytonema   160
Hormothamnion   160
Anabaena   161
Nostoc   164
Nodularia   166

Table 2. List of Species from the Older Literature Treated in this Work and the Valid Name of Each, *sensu* Drouet   168

GLOSSARY   173

BIBLIOGRAPHY   177

TAXONOMIC INDEX   189

SUBJECT INDEX   193

Introduction and Guide to the Marine Bluegreen Algae

INTRODUCTION

B luegreen algae are renowned for their ability to tolerate a wide range of environmental conditions, a characteristic of many primitive organisms. They probably represent, along with the photosynthetic and chemosynthetic bacteria, the survivors of the earliest photosynthetic plants. Their fossilized remains have been found in middle Precambrian rocks, and they may have appeared much earlier (Schopf 1970).

Their procaryotic cellular organization has restricted cell differentiation and confined them to microscopic size. Individual cells or plants of bluegreens are virtually invisible without magnification, and their identification usually requires the high-power objective lens of a microscope to see adequately the morphological characteristics that separate genera, species, and even families.

Biologists have been inclined to regard the taxonomy of bluegreen algae as inherently difficult and to invest them with an aura of mystique. For these reasons hair-splitting taxonomists, who have created scores of genera and hundreds of species based entirely upon trivial, environmentally controlled morphological characters, have gone unchallenged for nearly a century.

Another consequence of the taxonomic turmoil among the Cyanophyta, prior to the work of Dr. Francis Drouet, is that many of the major taxonomic guides to marine algae do not include them. Thus the student must try to find a local flora that does, or refer to the monographs of Drouet and Daily (1956) or Drouet (1968, 1973, 1978). The latter are technical taxonomic revisions not designed for the beginning student.

Thus there is a need for a general introduction and guide to identification of bluegreen algae that grow in the sea, based upon a sound and practical taxonomic system.

Keys, descriptions, and illustrations are included in this work to provide a means for determination *sensu* Drouet of bluegreen algae from all the world's oceans and estuaries. Drouet's taxonomic system for these plants is the most useful and logical of any currently available.

In addition, a means is presented in an appendix for determining the older names of most species of bluegreens recorded from a marine habitat, especially from the work of Gomont, Bornet and Flahault, Tilden, Fremy, Newton, Chapman, Cocke, and Desikachary. Many of these older names will continue to be useful to designate specific variants or ecophenes of the species Drouet has indicated as the valid ones.

Selection of species treated in this work has been, of necessity, somewhat arbitrary. Many species grow in both fresh and salt water. A few

2

may be restricted to estuaries in the marine habitat, but are more characteristic of fresh water. Species that may be brought into estuaries or the sea by river discharge or local run-off, but do not become established in estuaries, are omitted. With these criteria for "marine" bluegreen algae, there is a total of 34 species *sensu* Drouet treated in this work, 14 coccoid and 20 filamentous.

Dr. Francis Drouet is currently at work on his revision of the classification of the bluegreen algae family Stigonemataceae. Thus treatment of that family in this work is from the older literature, but only two species are recognized as marine.

All of Drouet's species of bluegreen algae recorded from marine habitats appear to be cosmopolitan in their distribution, in relation to their temperature and salinity ranges, except one. *Entophysalis endophytica*, a coccoid species, is known at present only in the Pacific Ocean.

Until the 1970s it was thought that the bluegreen algae were the only procaryotes containing chlorophyll *a*. In 1976 Lewin discovered a new phylum of procaryote plants, the Prochlorophyta, having the same chlorophylls as the green algae, *a* and *b*. Those known so far are unicellular and live symbiotically in didemnid ascidians. They have been found only in the Pacific Ocean on the Great Barrier Reef, at Eniwetok, and off Baja California. Dr. Lewin has expressed the opinion that they are ancestral to all green plants and that they constitute a missing link, the existence of which could have been predicted. The first species of the group was bluegreen and named *Synechocystis didemni* Lewin and Cheng (1975). It is now known as *Prochloron didemni* (Lewin 1977).

CLASS MXYOPHYCEAE, THE BLUEGREEN ALGAE

Of all the algae, the bluegreens are the most distinctive. In fact, they are so distinctive that bacteriologists call them "bluegreen bacteria," and their logic is really not open to question.

Not only are the bluegreen algae radically different from all other algae in their cytology, the characteristics most frequently emphasized, but they exhibit a primordial nonconformity in their distribution, ecology, reproduction, physiology, and biochemistry. They possess a multiplicity of characteristics that are unique among algae; they are an apple off another tree—the bacterial tree.

The distinctiveness of the bluegreen algae is derived from those characteristics that relate them to bacteria, characteristics that we associate with primitiveness, with early forms of life on a planet that has just reached the life-supporting stage in its evolution. These are the features of the procaryotic cell.

CLASSIFICATION

Recognition of the close relationships of bluegreen algae and bacteria was implemented in the seventh edition of *Bergey's Manual of Determinative Bacteriology* (1957) by placing them in the class Schizophyceae, and bacteria in the class Schizomycetes, the two comprising the division Protophyta of the Plant Kingdom.

The Editor-Trustees of the Eighth Edition of *Bergey's Manual* (1974) now place bluegreens and bacteria in the kingdom Procaryotae with the former in the division Cyanobacteria and the latter in the division Bacteria. They were not fully satisfied with this treatment but were "not prepared to subdivide the Procaryotae further at this time." R. G. E. Murray, however, suggested in the introduction that a future classification may look something like this:

Kingdom Procaryotae

       division.   Phototrophic procaryotes
        class I.   Blue-green photobacteria
       class II.   Red photobacteria
      class III.   Green photobacteria

     division II.   Procaryotes indifferent to light (Three classes)

This tentative classification suggests that the three major types of photosynthetic procaryotic organisms may receive equal treatment in the future.

The classification used in this work is the more traditional one, not because it is regarded as preferable, but only because it is so widespread in the literature and the classification with bacteria needs refinement.

The classification used here is based upon systems that first appeared in the 1930s and have since been widely used and occasionally modified. In this classification, bluegreen algae are still in the plant kingdom as the division or phylum Cyanophyta with one class, Myxophyceae, and two orders, Coccogonales and Hormogonales.

TAXONOMY

Dr. Francis Drouet, in his revisions of the classification and taxonomy of bluegreen algae, a work that spans 30 years, has removed the chaos from this group, chaos that resulted from the use of incredible trivia in separating species, genera, families, and even orders; morphological trivia that varies with the environment, but in the absence of knowledge of this fact, has been accepted as sound science during the decades and has led to the familiar taxonomic imbroglio of the bluegreens.

Drouet believes, as a result of culture work and the study of thousands of collections from all parts of the world, that characteristics of the sheath or gelatinous matrix of bluegreen algae are sufficiently influenced by the environment as to be taxonomically unreliable at the species level or above. Accordingly, he has based his classification upon cell or trichome characters.

Science is conservative. A radical revision like that of Drouet will not be generally accepted for a long time, even in the absence of good reasons for rejecting it. Lewin (1974, p. 29–30), in his consideration of this problem, failed to come up with one. He invoked some sort of general principle that "this is surely carrying 'lumping' too far," And he feels that we owe something to the bluegreen taxonomists of the past because of "the many tedious man-years of labour" that established "many apparently useful specific distinctions." The distinctions are useful, but most of them are not species.

Drouet's revisions are not the last word in the classification of the bluegreen algae. But they are a starting place. Since his classification is based upon morphology and since the bluegreen algae apparently do not exhibit enough in the way of stable morphological characteristics upon which to base a complete taxonomy, then we shall have to include some of the less convenient biochemical or physiological characters to establish sound species, genera, and families. It seems logical that eventually we shall base all bluegreen species upon cultures. The work of Baker and Bold (1970) and Stanier et al. (1971) is a good step in that direction.

Our concept of what constitutes a species of bacteria, a species with phylogenetic significance, is the poorest species concept applied to any category of living things. We have only the haziest ideas as to what constitutes evolution among bacteria, which characteristics that separate them are significant and which are not, which characteristics indicate phylogenetic relationships and which do not. Taxonomy in the lower categories is simply one of convenience, a product of a desperate need to separate bacteria into recognizable species and genera, even though it has to be done without a phylogenetic basis at the lower levels.

Bluegreen algae are bacteria. No one can deny that if "bacteria" is defined to include all procaryotic organisms. Furthermore, there is only one workable way to distinguish bluegreen algae from other bacteria, and that is their production of chlorophyll *a*. (So, there are no colorless bluegreen algae.) The Cyanophyta, however, are unequivocally set apart from other bacteria by their distribution, role, and conspicuousness in the environment. That is why they were classified with algae, aside from their chlorophyll, and that is why they will continue to be so classified, at least artifically, in the future.

Eventually we will have a taxonomic system that combines morphology, as far as it can be used, with whatever bacteriological-type characters are needed, based upon reference cultures. Until that time, the best we can do is to use Drouet's carefully devised taxonomy plus older names for the ecotypes or ecophenes that have been described in the past. In this way it is possible to indicate which variants of Drouet's "lumping" we have in hand.

This little book is designed to do just that for the bluegreen algae inhabiting the world's oceans.

GROWTH FORM AND MORPHOLOGY

Bluegreen algae are basically microscopic, although large colonies or plant masses may be quite conspicuous. Some species form spheres or cushions, and others produce closely adherent black strata or a low, dense turf. On soft bottoms of protected bays or estuaries they may occur as extensive mats, skeins, or gelatinous masses. Many are epiphytic upon larger algae or sea grasses where they produce microscopic patches or tufts, or expanding growths that may cover the host. In color, the plant masses range from yellowish brown to greenish black.

There are two principal groups of the bluegreens, coccoid and filamentous plants. Coccoid species range in growth form from single cells to colonies or masses of various sizes and shapes. In some colonies the cells are arranged in rows resulting in a flat plate, or they may be radially arranged in a spherical colony. In most, however, the cells are randomly distributed as a result of cell division in any plane. Colony size usually depends upon the effect of environmental conditions that promote fragmentation, but some species have inherent tendencies to fragment early and to produce few-celled colonies, whereas others tend to form many-celled masses.

Some colonies are loosely attached to the substrata without polarity. Others are firmly attached and have a distinct base and apex, often with basal parts that penetrate the substratum. All are enclosed in a gelatinous sheath that varies in consistency and thickness with environmental conditions.

Filamentous bluegreens produce a row of cells referred to as a trichome, a result of cell division in one plane and failure of the cells to secrete sheath material between the cells, or in the plane of division. The trichomes may have a firm, conspicuous sheath, or the secreted gelatinous material may be so diffluent that it does not accumulate, or so thin and colorless as to appear absent. Where an obvious sheath is present, the trichome and sheath together are referred to as a filament. The trichome is either loosely adherent or nonadherent to the sheath, and in many species trichomes or segments of trichomes frequently move out of the sheath and secrete a new one. Sheath secretion seems to be a constant process, but the rate, the degree of accumulation, and its consistency, are apparently a matter of environmental conditions.

With coccoid bluegreens, sheath material is secreted from all cell wall surfaces so that the cells are soon separated after cell division. The sheath usually holds the cells together in the form of a colony.

With filamentous species, the trichome can be regarded as a primitive form of multicellular plant despite the fact that there is little or no cell differentiation. With end walls of cells in contact, diffusion of substances from cell to cell may occur, or there may be movement of substances through ultramicroscopical cytoplasmic connections. Some filamentous species have modified apical or tip cells and some produce spores or slightly modified vegetative cells that function like spores, and some produce specialized cells known as heterocysts.

Attachment to the substratum is usually by the sheath, a result of its adhesive properties.

CYTOLOGY

During the past two decades much progress has been made in elucidating and interpreting the fine structure of cells of bluegreen algae. Certain new tools and procedures, including the electron microscope and its improvements, thin-section techniques, and the new precision microtomes have been of major importance.

Nucleoplasm and Chromoplasm

It has been traditional to regard the bluegreen cell as divided internally into two regions, an outer zone in which the pigments or photosynthetic lamellae are localized, the chromoplasm, and an inner zone in which the nucleic acids and nucleoproteins are localized, the nucleoplasm. Electron microscopic studies have shown that this concept may be somewhat misleading. Not only is there a transitional zone between these two apparent regions, as seen with a light microscope, but there is much variation that is often environmentally induced. In some cells it is impossible to distinguish two different areas. Pigments, although usually localized in outer areas, may extend through the center of the cell. Nucleoproteins, although typically concentrated in the center, may extend to the periphery.

DNA and RNA

Deoxyribonucleic acid (DNA) has been reported to occur in two forms in bluegreen cells, as anastamosing fibrils and as elongated rods. Neither is a chromosome, as a single one appears to be a complete genome (Fuhs 1969). The opinion has been expressed that there are 1, 2, 4, or 8 genomes per cell (Fuhs 1958, Hagedorn 1961). Hair cells of *Calothrix* contain

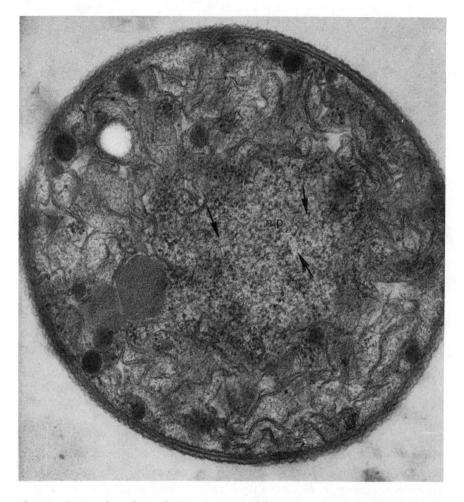

Plate 1. Section through a cell of *Anabaina variabilis* Kutzing showing the central location of the nucleoplasm and its numerous fine fibrils, the outer or peripheral location of the chromoplasm, the cell membranes, and cyanophycin granules in the chromoplasm (dark spheres). Electron micrograph courtesy Dr. Lee V. Leak, College of Medicine, Howard University, Washington, D.C., and with permission of Academic Press, New York.

approximately one-fifth as much DNA as the full-sized cells of the trichome.

Ribonucleic acid (RNA) is present in the bluegreen cell in the form of ribosomes, the protein-synthesizing organelles. They are distributed throughout the cell but are more concentrated in the nucleoplasm. Like those in bacteria they have a sedimentation coefficient of approximately 70, and they dissociate into 50s and 30s subunits.

Genome Replication

It is not yet known by what mechanism the genome units of bluegreen algae are replicated so that a full genetic equivalent is passed on to daughter cells. The extensive distribution of DNA within a cell suggests that even those species in which cell division produces daughter cells of unequal size, or a cell divides into many small cells ("endospores"), each daughter cell receives numerous genomes.

Photosynthetic Pigments

Photosynthetic pigments in bluegreen algae are localized in thylakoids, a system of flattened sacs bounded by a single membrane. Thylakoids tend to be concentrated at the periphery but may extend throughout the cell; the arrangement is influenced by light conditions (Land and Rae 1967, Lang and Whitton 1973). They are usually oriented parallel to the cell wall, but in the Nostocaceae the arrangement is often irregular.

Not only is chlorophyll *a*, the only chlorophyll in bluegreens, located in the thylakoids, but the three biliprotein pigments are also found on the lamellae of the thylakoids in the form of granules (phycobilisomes) attached to the outer face (Edwards and Gantt 1971, Gray et al. 1973, Evans and Allen 1973, Wildman and Bowen 1974). It has been shown that the amount of thylakoid material within a cell is proportional to the chlorophyll content (Allen 1968).

The phycobilin pigments, C-phycoerythrin (red), C-phycocyanin (blue), and allophycocyanin (blue), initiate the conversion of light energy to chemical energy through their susceptibility to ionization by certain wavelengths of visible light to which chlorophyll is not as sensitive. The energy thus absorbed by the phycobilins is passed on to chlorophyll, and chlorophyll carries on phosphorylation. Under some light conditions, absorption of light energy may take place to a greater extent by the phycobilins than by chlorophyll *a* in bluegreen algae.

The relative abundance of the various pigments in the bluegreens is profoundly influenced by the environment, especially light intensity, quality, and regime. Consequently the pigments will vary with time of year. Complementary color adaptation, the much-discussed phenomenon in which algae produce a preponderance of the particular pigment most efficient in absorbing the prevailing quality of light, occurs only under relatively low light intensities. Since there are a variety of environmental factors that influence the relative and absolute amounts of the pigments, complementary color adaptation may be eclipsed by other responses. Nitrogen deficiency may inhibit formation of the phycobilins more than it does chlorophyll; molybdenum and sodium deficiencies also inhibit synthesis of phycobilins. Green light favors the formation of phycoerythrin, and red light favors formation of the phycocyanins. White light of low intensity favors a high phycobilin concentration. Some bluegreen algae form phycocyanin only at high light intensity, and some apparently do not form phycoerythrin under any circumstances.

Bennett and Bogorad (1973), using a filamentous bluegreen alga referred to as *Fremyella diplosiphon*, found that C-phycoerythrin is not produced in culture under red light but it is formed under fluorescent light. In this species, it appears to be a photoinducible pigment.

Phycobilin pigments are closely associated with proteins from which they are not easily separated without alteration. Consequently most studies of them have been with biliproteins rather than with the phycobilins themselves. There is much yet to be learned about these pigment systems.

Carotinoids

Carotinoid pigments consist of two groups, the carotenes, which are oxygen-free hydrocarbons, and the xanthophylls, which are derived from the carotenes by oxidation. They are yellow, orange, or red. A variety of functions has been attributed to the carotinoids, and it appears that some have more than one important function. They may absorb light energy of wavelengths not readily absorbed by chlorophyll and then transfer electrons to chlorophyll. They serve to inhibit destructive photooxidation of chlorophyll and of the cell itself, at least in certain bacteria. They serve a coenzyme function as a precursor of vitamin A.

B-carotene is the most abundant carotene in the bluegreen algae and in most other photosynthetic plants. Myxoxanthin is one of the most abundant xanthophylls. At least 19 have been reported from the bluegreens (Prescott 1968). The carotinoids of *Spirulina platensis* have been studied by Tanaka et al. (1974).

As a result of the variety of pigments found in bluegreen algae and the variation in relative abundance in response to environmental conditions, the color of individual plants, or of a plant mass, potentially runs the gamut of the spectrum. Usually they are bluish green or green, but purple, maroon, brown, yellow, yellow-brown, or black are commonly encountered. Shades of red are most often found in plants from relatively deep water.

Extracellular pigments that develop in the sheath often contribute to the gross color of the plants. Sheath pigments may be yellow, brown, blue, violet, or red. Color of the plants has been used as a taxonomic species character, as have many other trivial differences, regardless of the role of the environment.

PHOTOSYNTHESIS

Bluegreen algae, like other algae and land plants, carry on photosynthesis by means of the two primary photoreactions I and II and the two pigment systems, photosystems I and II. Govindjee and Braun (1974) discuss the details of these light-energy-generated mechanisms.

Chlorophyll a has two major light absorption peaks at 430 nm (red) and 662 nm (blue), with four lesser peaks between these two. C-phycocyanin exhibits absorption peaks or shoulders at approximately 275 and 365 nm in the ultraviolet and a major shoulder at 615–620 nm (yellow). Allophycocyanin has a peak around 650 nm and considerable absorption around 620 nm. Phycoerythrin from the marine plankton bluegreen *Trichodesmium* (*Oscillatoria erythraea*) has triple peaks, according to Shimura and Fujita (1975) at 495–500, 543–547 and 557–565 nm. They indicate that it is more like the R-phycoerythrin of the red algae.

In addition to photosynthesis initiated by chlorophyll a, characterized by oxygen evolution from splitting water, *Oscillatoria limnetica* can switch to a bacteriallike photosynthesis in which H_2S is converted to sulfur (Cohen et al. 1975). *Anacystis nidulans* 1401 can utilize thiosulfate as an electron donor until all available is oxidized to sulfate, but only in the light. Under light-limited conditions that are not conducive to photosynthesis by chlorophyll a, this bluegreen can grow faster in the presence of thiosulfate than it can without it (Utkilen 1976). The question remains, how widespread is this ability among bluegreens?

Inhibition of photosynthesis by aeration was observed by Weller et al. (1975) in a species of *Phormidium* that forms conical growth masses in hot springs. They expressed the opinion that two phenomena are involved, the inhibition of photosynthesis by oxygen and the stimulation of carbon dioxide fixation by reducing agents, especially sulfide. The sulfide, however, was not being used instead of water as a source of hydrogen for photosynthesis. Within the conical growth masses of this alga, some of the filaments were in a favorable microenvironment for photosynthesis.

FOOD SOURCES

Bluegreen algae in general are obligate photoautotrophs; that is, they depend upon light as a source of energy and upon carbon dioxide as a carbon source. Most species will not grow in the dark on organic compounds.

Some degree of heterotrophic growth does occur, however, since the addition of sugars increases the growth rate of some species in the light. Some will assimilate acetate in the light but very little in the dark. A number of bluegreen algae can utilize egg albumin, serum albumin, peptone, and casein, probably through extracellular proteolysis. Many species can utilize various amino acids.

One bluegreen capable of growing heterotrophically (*Aphanocapsa* 6714) was shown to produce proteins, nucleic acids, and lipids at 10 percent the rate of photosynthesizing cells, but glycogen (myxophycean starch) was produced at approximately 50 percent the rate. After prolonged heterotrophic growth, myxophycean starch was the principal constituent of the cells, whereas in photosynthetically grown cells protein was the major constituent (Pelroy et al. 1976).

Wolk and Shaffer (1976), using a cultured strain of *Anabaina variabilis*, have shown that clones can be obtained that grow well in the dark on N-free media containing fructose. The dark-grown cells double every 28 hours, and every 9 hours in the light.

Plectonema boryanum is another species that has been grown successfully in the dark (White and Shilo 1975) with glucose, ribose, sucrose, mannitol, maltose, or fructose. The growth rate, however, is much more rapid in the light.

MUTATION

Mutations among bluegreen algae, both spontaneous and induced, are well known and seem to be accelerated by the same agents as are active with other organisms. Bluegreens seem to be more resistant to ultraviolet radiation than bacteria.

Ingram et al. (1972) treated the coccoid bluegreen, *Agmenellum quadruplicatum*, with the mutagen 1-methyl-3-nitro-1-nitrosoguanidine and obtained a stable tryptophane-requiring auxotroph. Mutants of this type are useful in analysis of gene-enzyme relationships. A cell division mutant was obtained by Ingram and Fisher (1973) from *A. quadruplicatum* that produced serpentine filaments, the length of which were temperature dependent. Another mutant from the same source produced septate filaments, a result of impairment of the cell separation process. Such mutants have resulted in a better understanding of cell division and growth, and they have indicated that the two processes are independent (Ingram and Fisher 1972; Ingram et al. 1973). Other workers have produced a variety of nutritional mutants by treatment with various mutagenic agents.

GENETIC RECOMBINATION

Although sexual reproduction, characteristic of eucaryotic organisms, does not occur in the *Procaryotae*, there is a mechanism for the transfer of genetic material from one cell to another. This phenomenon has been known for bacteria for two decades or more, but confirmation of its occurrence among bluegreen algae is more recent.

The earliest evidence of recombination in bluegreens was found in the easily cultured and widely used coccoid traditionally known as *Anacystis nidulans*. Two strains were obtained by selection of apparent mutants, one with an extremely high resistance to penicillin and another with resistance to streptomycin. The wild type or normal is susceptible to both of these antibiotics.

When cells of the two were mixed, a new strain appeared having resistance to both antibiotics (Kumar 1962, 1964). This circumstantial evidence for recombination stimulated further research. Kumar's work was confirmed with the same organism by Bazin (1968) and Shestakov and Khyen (1970). Evidence of genetic transfer was found in *Cylindrospermum majus* by Singh and Sinha (1965).

A mutant of *Nostoc muscorum* that was unable to fix nitrogen was obtained by Stewart and Singh (1975) and from it a strain resistant to 1000 μg/ml^{-1} streptomycin. This strain was mixed with the normal nitrogen-fixing, streptomycin-sensitive wild type. Later, nitrogen-fixing, streptomycin-resistant strains were obtained at a frequency of 4.6 in 10^5 colonies. Spontaneous mutation of each parent type growing alone was 1 in 10^7 or fewer. Since the frequency of production of the new strain in the mixture of "parent" types was approximately 400 times higher than in either strain growing alone, the difference was attributed to transfer of the nitrogen-fixing gene.

The mechanism of genetic transfer between bluegreen cells is not yet known. It may be effected by conjugation of donor and recipient cell as has been shown for bacteria, or exogenous DNA may be taken into a cell.

Genetic mapping of *Anacystis nidulans* has been undertaken by Asato and Folsome (1970) and Delaney and Carr (1975).

NITROGEN METABOLISM

Bluegreen algae can utilize the usual inorganic nitrogen sources assimilated by land plants. Ammonia is somewhat more readily used than nitrite or nitrate. They also can obtain their nitrogen requirements from a variety of organic sources such as urea, many amino acids, and casein hydrolysates.

A study of the available nitrogen sources for *Agmenellum quadruplicatum* by Kapp et al. (1975) may be representative of many bluegreens. They found that most amino acids were acceptable but that the growth rate varied widely among them. Cystine, methionine, and tyrosine did not support growth. Urea, uric acid, adenine, hypoxanthine, and xanthine were utilized. Of all nitrogen sources studied, however, NH_4Cl was best, followed closely by $NaNO_2$ and $NaNO_3$. The presence of these inorganic forms of nitrogen in the environment tends to inhibit utilization of organic forms. The culture of *Agmenellum* used in these experiments was said to be incapable of fixing nitrogen.

STORAGE PRODUCTS

Four kinds of reserve materials are found in bluegreen algae, carbohydrates, phosphate, fixed nitrogen, and lipids.

Myxophycean Starch

The principal food reserve is myxophycean starch, a glucan. This polysaccharide has been regarded as glycogen as it exhibits most of the reactions of animal starch. It differs, however, in that it cannot be fixed with permanganate. It appears to be similar to floridean starch of the red algae but the chain consists of 23 to 26 glucose units in contrast to an average of 15 in floridean starch (Meeuse 1962). It stains lightly with iodine, giving a reddish-brown or maroon color. It occurs principally as submicroscopic granules in the chromoplasm between the photosynthetic lamellae but it is also present in the nucleoplasm. It disappears after approximately 4 days of darkness and it does not accumulate in the presence of excess available nitrogen.

Polyphosphate Granules

Another nutrient reserve of bluegreen algae, referred to in the past as volutin or metachromatic granules, occurs as crystals of acid-soluble polyphosphates. These granules are spherical or oval, up to several microns in diameter, and probably occur at times in all bluegreen algae. They usually form in the centroplasm. Plants grown in phosphate-poor media lose them, but they form rapidly when phosphate is abundant in the medium or in the environment (Jensen 1968, 1969: Jensen and Sicko 1974: Sicko-Goad and Jensen 1976). That they consist of phosphate has been confirmed by Kessel (1977) by means of transmission electron microscopy and X-ray microanalysis.

Cyanophycin Granules

A type of granule that apparently is found only in bluegreen algae is known as cyanophycin. The accumulation of these granules on the cross walls of some filamentous species has been used as a principal taxonomic character to separate the genera *Microcoleus* and *Arthrospira* from other genera in the family Oscillatoriaceae by Drouet (1968).

Cyanophycin granules are visible with a light microscope without staining, and apparently occur in all bluegreen genera except *Spirulina*. They tend to disappear when cultures are kept in the dark, are soluble in mineral acids, are broken down by the enzyme pepsin, stained by acetocarmine, blackened by osmium tetroxide, and give a positive test for argenine. Simon (1971) indicates that they are a high molecular weight copolymer of argenine and aspartic acid. They apparently serve as a form of fixed nitrogen reserve (Simon 1973). Argenine and aspartate have been shown to be good sources of fixed nitrogen for *Agmenellum quadruplicatum* (Kapp et al. 1975).

Lipids

Bluegreen algae also store oil droplets in small quantity. Two sterols, cholesterol and *b*-sitosterol, and their various derivatives have been isolated (Reitz and Hamilton 1978, Nadal 1971), and others have been observed but not identified.

NITROGEN FIXATION

Many species of bluegreen algae are now known to be able to fix nitrogen. The fact that they do so without depending upon an external source of carbohydrate has led to the opinion that they may contribute more fixed nitrogen to the soil and the sea than do nitrogen-fixing bacteria. Bluegreen algae that fix nitrogen apparently have the simplest external nutritional requirements of any living things.

The newer techniques of detecting nitrogen fixation, the use of the isotope $^{15}N_2$, the acetylene reduction assay, the extraction of nitrogenase, the use of axenic cultures, have often confirmed older reports that had not been fully accepted. It seems logical to suspect that all species can do so to some extent under some circumstances. Only the more rapid rates of nitrogen fixation can be detected.

Since all algae lose organic nitrogen-containing compounds to the surrounding water, data on the rate of nitrogen-fixation may be low, as fixed nitrogen lost by diffusion from the cell is usually not taken into account. In algae grown in culture, there are data to suggest that this loss may be as much as 60 percent of the inorganic nitrogen compounds that are incorporated into amino acids (Fogg 1952).

In 1889 Frank of Germany first published evidence of nitrogen fixation by bluegreen algae. It was not until 1928, however, that convincing evidence was presented by Drewes. Since that time many papers on the subject have appeared and a long list of bluegreen nitrogen-fixing species has resulted.

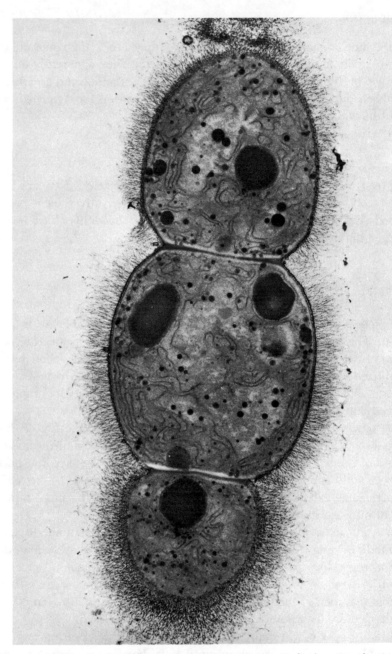

Plate 2. Three cells of a filament of *Anabaina sp.* in longitudinal section showing fine filaments that extend from the cell wall surface (outer convoluted membrane) into the mucilaginous sheath. The four large inclusions in these cells are cyanophycin granules. Electron micrograph courtesy Dr. Lee V. Leak, College of Medicine, Howard University, Washington, D.C., and with permission of Academic Press, New York.

Bluegreen algae are the only photosynthetic oxygen-producing plants capable of nitrogen fixation. Those in which this ability has been confirmed can be divided into two groups: species with heterocysts and those without.

The Heterocyst Species

Heterocysts have been shown to be the site of nitrogen fixation, or most of it, in those species that produce them (Van Gorkom and Donze 1971). Until this significant discovery, the function of the peculiar specialized cell of certain filamentous species known as a heterocyst had been a long-standing enigma (Fritsch 1951).

Heterocysts lack photosystem II (Donze et al. 1972), perhaps because they have no carboxysomes. Carbon compounds produced by photosynthesis in vegetative cells move into the heterocysts and supply energy essential to the production and activity of nitrogenase. In the absence of photosystem II, heterocysts maintain a sufficiently reduced internal environment to permit the nitrogenase enzyme system to function when the vegetative cells of the plant are carrying on photosynthesis and producing oxygen. It is believed, however, that vegetative cells may also carry on nitrogen fixation when they are under conditions of reduced oxygen tension, especially in the light.

Thiobacillus is a chemoautotrophic bacterium that contains characteristic cell inclusions, as seen by use of the electron microscope, known as carboxysomes. These bodies contain an enzyme essential to the photosynthetic (or chemosynthetic) process, ribulose-1-5-diphosphate carboxylase (Shively et al. 1973, Holt et al. 1974). Polyhedral bodies closely resembling carboxysomes have been found in many bluegreen algae (Stewart and Codd 1975), but only in vegetative cells and spores, not in heterocysts. It is believed that the reason heterocysts do not fix carbon photosynthetically is because they lack the enzyme contained in the polyhedral bodies (Winkenbach and Wolk 1973), and this in turn results in the absence of photosystem II.

Nonheterocyst Species

Nitrogen fixation by species of bluegreen algae that do not produce heterocysts has been reliably reported since 1964 (Dugdale et al. 1964, Goering et al. 1966, Taylor et al. 1973). In these reports, however, the possibility that nitrogen-fixing bacteria were responsible was not eliminated, as all dealt with the abundant marine planktonic bluegreen, *Oscillatoria erythraea*, formerly known under four species of the genus *Trichodesmium*. This species has not yet been grown in culture for an indefinite period.

Another source of skepticism was the fact that nitrogen fixation is most rapid in the light (Fay 1970). Although photosynthesis generates reducing power and adenosine triphosphate (ATP), both of which are essential to nitrogen fixation, at the same time the production of oxygen as a by-product of photosynthesis inhibits nitrogen fixation.

Wyatt and Silvey (1969) showed that a strain of the coccoid bluegreen known as *Gloeocapsa* produced nitrogenase, although its activity was light-dependent under aerobic conditions. Its rate of nitrogen fixation was said to be equal to that of heterocyst-producing species. Gallon et al. (1974) working with two so-called strains of *Gloeocapsa* found that in young cultures in a nitrogen-free medium photosynthetic oxygen evolution was low and nitrogenase activity at its peak. As the culture aged, there was an increase in photosynthetic pigments and photosynthesis, and nitrogen fixation fell to a low level. Reserves of fixed nitrogen apparently accumulated while the culture was young. Since the two processes were somewhat separated in time, both occurred at high rates at different times. The culture conditions, however, were probably not typical of a natural environment.

An explanation for the rapid occurrence of nitrogen fixation in a species without heterocysts was suggested by Carpenter and Price (1976), working with the planktonic marine species *Oscillatoria erythraea*. They observed slightly differentiated cells in the central area of colonies or clusters of filaments having much-reduced pigmentation in comparison to cells on the outside or near the ends of the filaments. The centrally located cells did not take up $^{14}CO_2$ even under optimum conditions for photosynthesis. They apparently do not evolve O_2 and thus maintain a reducing environment conducive to the activity of nitrogenase in the light when relatively large quantities of ATP are available. At the same time, the cells on the periphery of the colony are carrying on photosynthesis. The centrally located cells may lack photosystem II, like heterocysts, although this has not yet been demonstrated.

The propensity of *O. erythraea* to form loose, spherical, or flattened, three-dimensional colonies is a characteristic that contributes significantly to its ability to thrive and to produce vast blooms in warm seas during seasons of the year when competition from other phytoplankton species is lowest. Periods of calm weather that permit maximum development of the colonies because of reduced water turbulence also promote the best development of *O. erythraea*.

There are other species of both filamentous and coccoid bluegreens that lack heterocysts that are known to fix nitrogen, especially under conditions that are microaerobic. An example is *Plectonema boryanum*, a form of *Schizothrix calcicola* (Stewart and Lex 1970).

Bluegreen algae fix nitrogen in the dark also, but at a much-reduced rate.

THE CELL WALL

The cell wall of bluegreen algae is a four-layered structure between the plasmalemma or cytoplasmic membrane and the sheath material. It differs from the wall of all other algae and from that of land plants. It determines the shape of the cell, provides a degree of rigidity, and holds the cell contents intact during changes in osmotic pressure of the surrounding medium. Apparently it does not contain cellulose (except in the case of heterocysts), chitin, or pectin.

Just outside the plasmalemma is an electron-transparent layer approximately 10 nm thick, although there is some doubt about this part of the wall as some electron microscope preparations do not show it (Jost 1965, Halfen and Castenholz 1971). It may be an artifact.

The next layer makes up the major portion of the wall and consists of a mucopolymer of homogeneous composition that seems to provide rigidity. It is sensitive to lysozyme, an enzyme that digests the walls of bacteria and bluegreen algae and causes lysis of the cells. There are pores or thin spots in this wall layer in rather regular distribution. The sheath material may be secreted through these.

The third layer is a thin one of approximately 12–14 nm and is characterized by an abundance of fibrils with a 60° helical orientation (Halfen and Castenholz 1971). The outer layer is also thin (8–9 nm) and is continuous over the trichome or cell but does not occur in cross walls.

A major portion of the wall layers consist principally of glycoaminopeptides (Salton 1964) of which muramic and diaminopimilic acids are characteristic constituents. These two acids are known only in the walls of bacteria and bluegreen algae and account for their dissolution by lysozyme and their sensitivity to penicillin in the sense that penicillin interferes with their formation. The glycoaminopeptides form the fibrils of the third layer of the wall (the second layer from the outside).

Shrinkage of the protoplast in a hypertonic solution may result in a plasmolytic form that is concave. Or the protoplast may separate from the cell wall at intervals and only for a brief period followed by osmotic regulation and deplasmolysis. The protoplast is highly viscous and the wall quite elastic.

Lysis of Bacteria

Ingram (1973) has shown that at least some bluegreen algae produce enzymes that lyse bacteria by dissolution of the cell wall. He obtained lytic substances from *Agmenellum quadruplicatum*, *Anacystis nidulans*, and a species of *Nostoc* that were active against the gram positive coccus, *Micrococcus lysodeikticus*.

Lytic enzymes may serve to reduce competition in the natural environment, especially for planktonic species of bluegreens. Separation of cells after division is a function attributed to lytic enzymes produced by bacteria, one that may apply also to coccoid bluegreen algae.

There are bacteria that lyse bluegreen algae either by production of extracellular lysozyme or by attaching to the bluegreen cell (Shilo 1966, 1970; Stewart and Brown 1969; Daft and Stewart 1971). Apparently all these bacteria belong to the Myxobacteriales, principally the genus *Cytophaga*.

THE SHEATH

All bluegreen algae secrete a gelatinous material which, in most species, tends to accumulate around the cells or trichome in the form of an envelope or sheath. Coccoid species are thus held together to form colonies. In some filamentous species the sheath may function in a similar manner, as in the formation of *Nostoc* balls or in development of the firm, gelatinous hemispheres of the marine *Phormidium crosbyanum* ecophene of *Schizothrix calcicola*. In both cases, masses of trichomes are held together in a common gelatinous matrix. More commonly, the sheath material in filamentous species forms a thick coating or a tube through which motile trichomes move readily.

Rate of Production

Sheath production is a continuous process of bluegreen cells, and those that remain within a sheath or matrix for an extended period will produce a thick one. Variations in the rate of sheath production or in the periodic fluctuation of an environmental factor may result in distinct laminations. There are circumstances that tend to reduce the rate of growth and cell division relatively more than the rate of sheath secretion, and vice versa.

Pigmentation of the Sheath

Under some environmental conditions the sheath may become pigmented, although it is ordinarily colorless and transparent. Ferric hydroxide or other iron or metallic salts may accumulate in the sheath. Xanthophyll or anthocyanin pigments originating within the cell may color the sheath, the production of which may be partly hereditary, partly environmental. A serine-threonine peptide that turns yellow-brown after excretion by the cells was reported by Walsby (1974) as responsible for sheath color. Parasitization by fungi, especially chytrids, sometimes discolors the sheath.

Chemical Nature and Physical Properties

It is strange that so little is known about the chemical nature of the extracellular hydrophilic colloids that form the sheath or matrix of bluegreen algae. Although polysaccharides are involved, it does not seem to be a pure polysaccharide. Mucopolysaccharides have been reported. The sheath may be of the same material as the continuous gelatinous layer of the cell wall or include some of this material in its structure.

In physical properties the sheath material of some bluegreen algae resembles the extracellular polysaccharides of the red algae, agar and carrageenan. If a sufficient quantity of one of these species is air dried and then autoclaved at 15 pounds pressure for 15 minutes in the proper volume of water, a colloidal solution will result that behaves as a thermally reversible gel when cooled. The polysaccharides of the red algae are sulfuric acid esters of linear galactans.

The sheath of *Anabaena flos-aquae* is reported to consist of fibrils 1–2 nm in diameter and oriented parallel to each other and to the cell surface. In other species or strains the fibrils have been found to be perpendicular to the cell surface (Leak 1967, Lang 1968).

Function

A principal function of the sheath appears to be the protection it affords those species that grow in the intertidal zone and are thus alternately exposed to the marine and terrestrial environments. At low tide these plants are exposed to a range of temperature from that reached by solar heating in summer of the rocks to which they may be attached to the subzero air temperatures of winter. They are also exposed to low humidity and consequent dessication and to rain or snow. Species that grow in fresh water or in soil may be able to survive long periods of dryness because of the sheath. On the other hand, if the function of the sheath is protection from conditions such as these, then it is difficult to explain why well-developed sheaths occur on marine species of bluegreens that are always submerged in an environment of relatively high stability.

Lange (1976) suggested that the polysaccharides of the sheath have the property of absorbing from the environment essential nutrients that are present in submarginal levels and thus make these nutrients readily available to the plants.

MOTILITY

Many species of bluegreen algae exhibit a creeping-rotating motility when the cells or trichomes are in contact with a solid or semisolid surface. Since bluegreen algae do not have flagella, the mechanism of this movement has been a baffling problem.

Several tentative explanations for this peculiar form of motility have been made. One of these postulates a directional secretion of polysaccharide through pores in the wall and the immediate hydration of the polysaccharide with a propelling effect (Frank et al. 1962). Another theory accounts for the movement by waves propagated along the cell walls, an idea discussed by Jarosch in Lewin (1962).

Halfen and Castenholz (1971) came to the conclusion that the gliding motility and rotation of bluegreens is a result of unidirectional waves in the fibrils in the layer of the cell wall that is second from the outside. The fibrils are arranged in parallel array in a 60° helical orientation. They transmit their coordinated waves of bending through the thin outer layer of the wall to the outside. This motility-producing force is impressed against the sheath or substrate.

These fibrils are 6–9 nm in diameter and appear to be the same as those making up bacterial flagella, which are 4–6 nm in diameter. Activity of similar fibrils is believed to account for the cytoplasmic streaming in the freshwater green algae *Nitella* and *Chara*, in the myxomycete, *Physarum*, and in the gliding motility of the myxobacteria, *Chondrococcus* and *Cytophaga*.

The rate of movement of bluegreen algae is influenced by temperature, viscosity of the external medium, pH, and other factors. Evidently ATP derived directly from oxidative phosphorylation is an essential energy source, since DNP inhibited movement both in light and in darkness.

Trichomes of *Oscillatoria princeps* were observed to travel 188 μm per rotation at a maximum speed of 11.1 μm per second. Since the trichome observed had a diameter of 35 μm a point on its surface would travel approximately 110 μm per revolution. It was calculated from these figures that a given point on the surface inscribes a helix with a pitch of 60° as the trichome moves.

Perhaps the puzzle of the movement mechanism of bluegreen algae has now been solved.

REPRODUCTION AND DISSEMINATION

Reproduction in bluegreen algae is strictly asexual. The basic mechanism is simple cell division or fission that ultimately leads to separation of the daughter cells or fragmentation of the colony or trichome and the transportation of cell, fragments of a colony, or trichome to another location. There are, however, some minor modifications that increase the efficiency of dissemination.

Endospores and Hormogonia

Among the Coccogonales are several genera and species in which larger vegetative cells will often undergo repeated divisions resulting in the formation of a few to many small cells enclosed within the original sheath. These cells have been called endospores if they are of medium size, nannocytes if they are numerous and small. They are produced in the sea by the genus *Entophysalis*. All sizes of these endospores may be produced by the same plant.

In the family Clastidiaceae the plants are solitary, erect, and attached at the base. They begin as elongate single cells surrounded by a sheath. Cell division in one plane at the top at right angles to the axis results in the formation of a terminal chain of cells. As these enlarge, they rupture the sheath and escape. They have been referred to as exospores, although they do not differ in any way except size from the basal cell of the plant. They are simply the result of unequal cell division. The Clastidiaceae occur in freshwater only.

Among filamentous genera there is a tendency for trichomes to fragment into short segments referred to as hormogonia. These segments, if they are within a sheath, usually move out and serve as a sort of propagulum. In some species, or under some circumstances, they are remarkably uniform in length, but in most the length varies from a few to many cells so that there is no distinction between a hormogonium and a purely vegetative trichome. Hormogonia may be produced as a result of secretion of sheath material between two cells of a trichome in the form of a convex-lenticular pad. The trichome breaks at this point and the end cells become rounded. Trichomes may be segmented into hormogonia also by separation at a heterocyst or by death of an occasional cell. Occasionally trichomes will fragment into single cells, and these have been termed "planococci." Their formation is often associated with adverse conditions.

Endospores, exospores, and hormogonia are so similar to vegetative cells or trichomes, as the case may be, that they hardly deserve a specific term. They are no more resistant to adverse conditions than vegetative cells or plants and their cell walls are not modified in any way.

Akinetes

The only distinct spores of the bluegreen algae are those produced in the families Nostocaceae and Scytonemataceae of the order Hormogonales. An akinete develops from a vegetative cell. The cell enlarges, becomes dense with food reserves and, in the final stages, the wall becomes thicker and the sheath may become continuous around the ends of the cell. The original vegetative cell wall also serves as the outer wall of the akinete. These spores have the ability to tolerate adverse conditions to a greater extent than do vegetative cells.

In some species akinetes are always adjacent to a heterocyst or terminal. In others they are randomly distributed. Usually they are single but in some species of *Nodularia* they occur in chains.

Akinetes are common among freshwater species of the families mentioned but uncommon in the sea, probably because of the stability of the marine environment. Dessicated akinetes in soil samples have survived for 70 years (Bristol-Roach 1919, 1920). Dried colonies or plant masses of vegetative material of some species have been known to survive almost as long in herbaria.

Akinete formation, at least in some freshwater species, is controlled by critical levels of available fixed nitrogen and carbohydrates (Tyagi 1974). In the presence of ammonium chloride or potassium nitrate, spore formation was inhibited. In the presence of glucose, all vegetative cells of *Anabaena doliolum* formed akinetes beginning with those farthest from heterocysts. Phosphate deficiency also stimulates sporulation.

HETEROCYSTS

The most distinctive modification of a vegetative cell among the bluegreen algae is that of the heterocyst. Like akinetes, they are produced only in the families Nostocaceae and Scytonemataceae of the Hormogonales. Heterocysts differ from vegetative cells principally by their greater wall thickness, their translucent nongranular contents, their lighter often yellowish color, the absence of biliprotein pigments of photosystem II, and the absence of polyhedral bodies or carboxysomes.

Development

Heterocysts develop from vegetative cells in response to certain environmental conditions. In the process three additional wall layers are formed on the outside of the vegetative cell wall (Lang 1965), an inner laminated layer of glycolipids, a thick homogeneous middle layer that is said to be cellulose (Granhall 1976), and an outer fibrous layer (Lang and Fay 1971). Polyhedral bodies, polyphosphate and cyanophycean granules disappear as the heterocyst develops. The thylakoids become rearranged in a honeycomb form near the end wall, but the ribosomes are unaffected. The chlorophyll content is slightly reduced, b-carotene becomes predominant, and the phycobilin pigments, phycoerythrin, phycocyanin, and allophycocyanin, disappear or become extremely dilute.

Plasmodesmata

Early in heterocyst development cytoplasmic connections become established with adjacent vegetative cells. As the heterocyst matures these

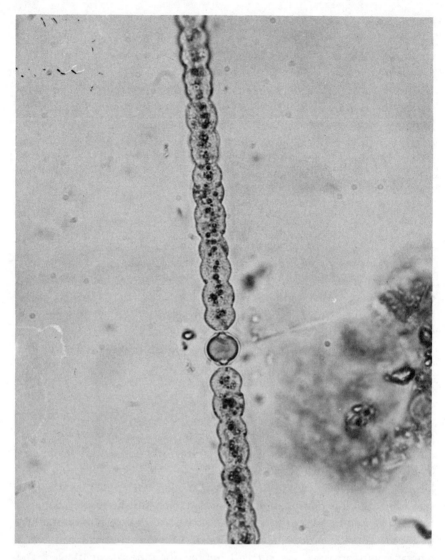

Plate 3. A filament of *Anabaina sp.* showing a mature heterocyst and polyphosphate bodies in the vegetative cells, revealed by the Ebel staining procedure. Microphotograph courtesy Dr. Thomas S. Jensen, Department of Biological Sciences, Herbert H. Lehman College, Bronx, N.Y., and Dr. C. C. Bowen, Department of Botany, Iowa State University, Ames, Iowa.

plasmodesmata develop gelatinous plugs, referred to as polar bodies, on the inside of the end walls, but microplasmodesmata remain. These plugs, along with the larger size and homogeneous-appearing contents of a heterocyst, are the distinguishing characteristics. Terminal heterocysts have only one gelatinous plug as they are in contact with only one vegetative cell. Heterocysts located at the junction of a branch in the Scytonemataceae have three gelatinous plugs.

Location

Heterocysts are usually solitary and may be terminal, intercalary, or both. In some species they occur in chains. Heterocyst formation, hence abundance, is inhibited by a relatively high availability of fixed nitrogen but is promoted by nitrogen starvation, insufficient phosphate, and high availability of an exogenous source of glucose or other assimilable carbohydrate.

Wall Resistance

The nature of the wall of mature heterocysts is responsible for the fact that lysozyme will not affect them, cyanophages do not attack them, and bacteria that lyse vegetative cells do not destroy heterocysts. When gas vacuoles form in vegetative cells of a trichome, the heterocysts are conspicuous for the absence of these vesicles. These properties of the heterocyst may be derived from the presence of cellulose in the wall (Granhall 1976).

Function

The role of the heterocyst has been a puzzle ever since its discovery (Fritsch 1951), and the problem has generated much speculation. It is now believed that its principal function is that of nitrogen fixation (Stewart et al. 1968). In this process it also influences the position and number of spores or akinetes formed by heterocyst-producing species.

GAS VACUOLES

Gas vacuoles have been known to occur in certain bacteria and bluegreen algae since the last century. Three German microbiologists in 1895 showed that these tiny refractive cell inclusions actually contained gas. They completely filled a bottle with a suspension of planktonic bluegreen algae containing gas vacuoles, placed a cork in the bottle, and produced a sudden increase in hydrostatic pressure by striking the cork with a mallet or hammer. After this treatment the algal cells slowly sank and, upon microscopic examination, they were seen to be devoid of gas vacuoles (Ahlborn 1895, Klebahn 1895, Strodtmann 1895). Under a vacuum, however, the gas vacuoles persisted, suggesting that they were something more than free bubbles.

Under the electron microscope, gas vacuoles, perhaps better referred to as gas vesicles, are seen as cylinders of rather uniform diameter but of various lengths with conical or rounded ends and bounded by a membrane approximately 2 nm thick (Bowen and Jensen 1965). The diameter of the cylinders is 70–75 nm and their length 200–1000 nm, with a mean length of approximately 370 nm. Walsby (1972), in an excellent review of current knowledge of gas vesicles, lists 13 "species" in eight "genera" in which these inclusions had been studied up to that time, many of which are planktonic.

Jones and Jost (1970) devised a procedure for separating gas vesicles from the cells, and found that the membrane is protein and has a distinct ribbed structure with the ribs 4.0 to 4.5 nm apart.

The gas vesicle membrane is highly permeable to gases, as is indicated by the fact that the application of hydrostatic pressure causes their instant collapse. The gas inside them apparently is air. They do not function as a storage reservoir of any kind of gas used in cell metabolism, as their contents are simply the gases dissolved in the surrounding medium. They cannot be inflated by the application of pressure. Their total volume ranges from 1 to 10 percent that of the cell.

Two possible functions have been attributed to them, the provision of buoyancy to the cells or plants, and the shielding of the thylakoids from strong light.

They are effective light scatterers. In *Anabaena flos-aquae* under low light intensity they are distributed throughout the cell with random orientation. Under high light intensity they move to the cell periphery and are radially oriented (Shear and Walsby 1975). When the light intensity is low they are numerous enough in *Anabaena* to make the cells buoyant, but under high light intensity the alga becomes heavier than the culture medium, probably because a high rate of photosynthesis increases the osmotic pressure inside the cells and this in turn causes collapse of some of

Plate 4. Gas vacuoles in *Nostoc muscorum* in longitudinal section as seen in an electron micrograph made by freeze etching. These are mature vacuoles in which the unique banding pattern is remarkably distinct. They are up to 1.5 μm in length and about 0.1 μm in diameter. Photograph courtesy Dr. J. Robert Waaland, Department of Botany, University of Washington, Seattle, and with permission of *Science* (A.A.A.S.).

31

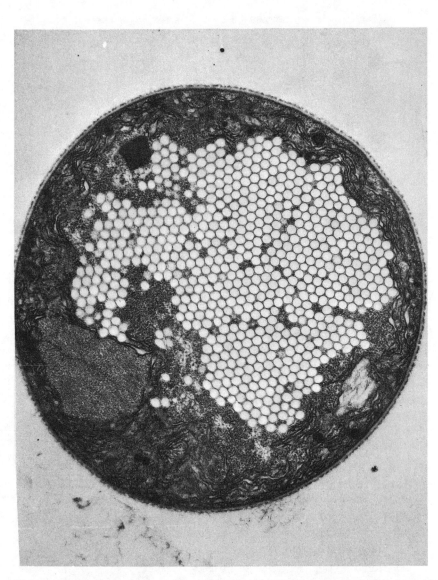

Plate 5. Gas vacuoles in *Nostoc coeruleum* in cross section as seen in an electron micrograph. Photograph courtesy of Dr. Thomas S. Jensen, Department of Biological Sciences, Herbert H. Lehman College, Bronx, N.Y., and Dr. C. C. Bowen, Department of Botany, Iowa State University, Ames, Iowa.

the gas vesicles. Present knowledge indicates that their function is buoyancy rather than light-shielding.

Among bluegreen algae that thrive in the sea, *Oscillatoria erythraea* is the species that seems always to have gas vesicles. They are at the periphery of the cell and are probably the reason this species is strictly planktonic either because they promote buoyancy or because they protect the photosynthetic system from sunlight that is too intense, or both. During the warmer months of the year *O. erythraea* forms vast areas of discolored water in temperate and tropical seas such as the Gulf of Mexico (Curl 1959) and the Sargasso Sea.

Van Baalen and Brown (1969) found structures in the center of cells of *O. erythraea* referred to as cylindrical bodies. There were one to three per cell, their diameter was 0.2–0.3 μm, the length variable. As seen in cross section, they contained two double-lamellar units that were similar or identical to the photosynthetic lamellae, forming two concentric rings, and a central core approximately 200 Å in diameter. Various degrees of dissociation of the cylindrical bodies that they observed led Van Baalen and Brown to postulate that they were the site of synthesis of the photosynthetic lamellae. Earlier, Pankratz and Bowen (1963) had noted cylindrical bodies in *O. retzii* (*Symploca muscorum*), a freshwater species.

Since it has not been possible to maintain *O. erythraea* in culture indefinitely, it has been difficult to study it. When collected from the sea, the cells tend to lyse within a few minutes. Van Baalen and Brown (1969) observed that lysis occurs following a rupture at the junction of the longitudinal and transverse walls.

A marine bacterium in which gas vacuoles are conspicuous is *Lamprocystis roseo-persicina*, a coccoid organism that produces patches of rose-colored sand in the intertidal zone of low-energy beaches where it loosely cements the sand particles together. It is often mistaken for a small coccoid bluegreen alga. The intertidal patches of *Lamprocystis* may be covered with a few millimeters of sand, as it carries on photosynthesis anaerobically, obtaining hydrogen from hydrogen sulfide rather than from water. The patches are sometimes exposed after windy weather has caused removal of the thin layer of sand under which it normally grows.

CYANOPHAGES

In 1963 Safferman and Morris announced the discovery of viruses that attack bluegreen algae, the cyanophages. Since that time, many have been reported that attack a number of species of bluegreens. Cyanophages are named in accordance with their known hosts, but this system is not entirely satisfactory because of the chaotic condition of the older bluegreen taxonomy, so much of which is based upon minor morphological characters that vary with the environment. Padan and Shilo (1973) have suggested that cyanophage serology may be in bluegreen taxonomy in the future.

Cyanophages are of the head-tail type. At least three groups are known, differing as to bluegreen host group and also in virion size, morphology, and other characteristics. They have a six-sided head and long or short tail. The nucleic acids of those known so far are double-stranded, linear DNA.

The time from adsorption of a virion onto the wall of the bluegreen cell until the cell bursts and many virions are released varies generally from 12 to 24 hours, and the number of new virions is usually from 100 to 350 per host cell. Bluegreen cells kept in the light after infection produce many more virions than those kept in the dark. In fact, one type of cyanophage fails to reproduce in the host cell if it is kept in the dark, as it is dependent upon the photosynthetic activity of the host (Safferman and Morris 1967).

Although most work on cyanophages has been with bluegreen algae from nonmarine sources (especially from sewage ponds), it seems highly likely that cyanophage activity is relatively as prevalent in the sea. The species most often observed as a host is *Schizothrix calcicola* (*sensu* Drouet 1968), one of the most abundant and widely distributed bluegreens both in freshwater and the sea.

DISTRIBUTION IN THE SEA

Although more abundant and in greater variety on land and in freshwater than in marine habitats, bluegreen algae are an important group among plants of the sea. Their greatest abundance in is the intertidal zone, but they also occur in significant quantities throughout most of the euphotic zone as benthic, attached plants and also in the plankton. Occasionally they are found in abundance in the aphotic zone.

Intertidal Zone

Wherever there are rocks, seawalls, or pilings, the upper intertidal zone is clearly marked by bluegreens. They form a "black zone" or upper intertidal band that is black or nearly so. In most habitats this band is almost exclusively *Calothrix crustacea*, but in some places, especially on creosoted pilings, the *Lyngbya confervoides* ecophene of *Microcoleus lyngbyaceus* is the colonizer. This species is tolerant of the cresols, phenols, and other toxic compounds that diffuse out of pressure-creosoted wood. Hence it is characteristic of harbors, estuaries, polluted, or brackish waters. It forms a skin on smooth surfaces, often in a pure stand.

The most luxuriant development of bluegreens in the intertidal zone occurs in tropical marine waters. At the high-tide line or even above the *Calothrix pilosa* form of *Scytonema hofmannii* forms a dense blackish turf approximately 1 cm thick on rocks and on the aerial roots of red and black mangroves. Somewhat lower down, but best developed in the upper half of the intertidal zone, is a variety of other species in many growth forms. Some form crusts, others produce plicate velvetlike growths, and some form gelatinous spheres, hemispheres, or layers. Their low, compact form and tenacious attachment enable them to thrive in wave beaten areas, but they also occur in quiet, protected waters. On exposed coasts the black zone may be above the high-tide line in the splash or spray zone.

Below Low Tide

Some of the same species that are predominant in the intertidal zone occur also below low tide to considerable depths but other species are found only in subtidal waters. In general, the bluegreens are not conspicuous in deeper waters. In a few habitats such as protected bays they may form an extensive mat over muddy or muddy-sand bottoms, or they may form coatings, patches, or entangled skeins upon and among seagrasses, especially during the warmest months.

Even if they are not evident, they may be present in abundance. Virtually all larger algae have numerous bluegreen epiphytes but they may not be detected until the host plants are examined under magnification. These bluegreen epiphytes are omnipresent to almost the greatest depths at which the host plants occur. Whether or not they derive nourishment from the host has not been determined, but they probably do to some extent in view of their ability to utilize a wide variety of organic compounds. Host specificity among marine bluegreens is rare.

Below the Euphotic Zone

Bluegreen algae have been reported many times in unexplained abundance from various depths below the compensation plane for photosynthetic organisms, usually as plankton. Apparently they are living heterotrophically or chemoautosynthetically and are facultatively photosynthetic.

Bernard (1963) reported a small *Nostoc* ("near *N. planctonicum* Poretzky") at depths of 1000 meters in the Mediterranean Sea and the Indian Ocean.

In the Plankton

Although a number of species of bluegreen algae are often found as plankton, apparently only one species is strictly planktonic. *Oscillatoria erythraea* is widespread in tropical and subtropical seas of the world as unbranched filaments in the form of bundles or fascicles held together by an invisible or diffluent mucopolysaccharide sheath. This species often forms vast blooms or patches of discolored water or a "red tide." The patches are sometimes red but more often they are yellowish or amber. The colonies are visible to the eye as tiny flakes, often concentrated at the surface during calm weather. They are the source of the name for the Red Sea. Sailors have often referred to this organism as "sea saw dust."

O. erythraea is reported to occur in greatest abundance during the warmest months of the year when the surface waters are lowest in nutrients (Hulburt et al. 1960). It is reported to be a nitrogen-fixer.

O. miniata, a planktonic form of *Microcoleus lyngbyaceus*, was formerly regarded as a strictly planktonic species. It occurs as single trichomes without a visible sheath, and has been reported from the Adriatic Sea, the Indian Ocean, the Caribbean Sea, and the Sargasso Sea.

Within Limestone

The most assiduous penetrators of limestone of all algae are the bluegreens. Any shell or piece of limestone in the intertidal zone or below that has a greenish tinge is very likely to have a growth of bluegreen filaments within to a depth of 1–5 mm, depending upon light penetration. If the shell or stone is broken the depth of penetration can be determined roughly by examining a cross section, but the plants themselves can be studied and identified only after a piece of the limestone has been decalcified in dilute hydrochloric acid. The acid treatment does not seriously alter morphology of the plants.

Stony corals usually have a dense growth of bluegreens that contribute organic matter to the coral polyps. The same species of bluegreens found in living corals are also found in limestone of nonliving origin, in dead corals, in the shells of living or nonliving molluscs, the calcareous tubes of polychaete annelids, tests of calcareous bryozoans, foraminiferans, and other invertebrates. These same bluegreens are also found growing on the surface of calcium carbonate and noncalcareous substrata. The growth form varies widely with the microhabitat and other environmental conditions with the result that some species have been described under many generic names and a host of species names (Drouet 1963).

Entophysalis of the family Chamaesiphonaceae produces strata or cushions of coccoid cells on the surface of a substratum. On limestone or wood these plants penetrate the substratum, and within the substratum they produce a pseudofilamentous growth, as might be expected. The filamentous endolithic growth form has been known under the genus *Hyella*, whereas the coccoid cells on the surface have been placed in a number of genera including *Dermocarpa, Xenococcus, Pleurocapsa*, and *Aphanocapsa* as well as *Entophysalis* as previously defined in the family Chroococcaceae.

Bluegreen algae that secrete organic acids and bore into limestone are important agents of the destruction of all forms of calcium carbonate in shallow waters, especially in warm seas (Purdy and Kornicker 1958). In a study of boring bluegreens at St. Croix, Virgin Islands, West Indies, Perkins and Tsentas (1976) placed pieces of calcium carbonate at many depths and localities from the intertidal zone to 30 m and analyzed them at intervals to determine the time required for bluegreens to invade new substrata, which species do so, and their relative abundance in relation to depth from 0 to 30 m. They found six species of limestone borers in the following order of abundance: *Mastigocoleus testarum, Plectonema terebrans, Calothrix* sp., *Hyella* species "A" and "B", and *Scytonema* sp. *Calothrix* and *Scytonema* apparently are newly reported as carbonate borers. Since these names are from the older literature, there appear to be five species *sensu Drouet* (1968, 1973) and Drouet and Daily (1956). Most of the carbonate pieces placed in the sea were bored into in less than 4 weeks time and were heavily infested in less than 6 months.

There are some bluegreens, on the other hand, that contribute to the formation of limestone by serving as nuclei for limestone precipitation. Calcareous concretions or ooliths in ancient marine deposits may have originated in part in this way.

TEMPERATURE AND SALINITY

Although temperature and salinity are factors of importance in the distribution and behavior of bluegreen algae, the influence of these two environmental features on bluegreens is much less than it is on other groups of algae. Most bluegreens are eurythermal and euryhaline. As a consequence, bluegreen algae are so widely distributed, especially if the taxonomy of Drouet is followed, that a local flora is useful for identification in almost any part of the world.

There are a few tropical species not found in temperate or cold areas, and there are a few cold or temperate species not found in the tropics. The majority of bluegreen species can grow equally well in fresh- or saltwater, but there are a few found only in freshwater and a few found only in the sea. Many species characteristic of freshwater range into brackish water, but do not occur permanently in the sea. Other species that are well adapted to the sea, also occur in brackish water but do not persist in fresh water.

In hot springs, some bluegreens can tolerate water temperature to 74°C. Other species grow in Arctic and Antarctic regions where the temperature is always below 0°C. Whereas those that occur in the sea rarely encounter water temperature above 32, or below -2°C, certain species that occur in the intertidal zone demonstrate wide tolerances of salinity and temperature.

The mile-long jetty flanking Aransas Pass at Port Aransas, Texas, is composed of huge blocks of granite. On top of the jetty are rough-hewn blocks that are wet with seawater during winter weather and by rains. There are numerous depressions that hold shallow pools of water and in these temporary pools is a crust or mat of *Calothrix crustacea*. During windy weather, the pools are seawater of a salinity ranging generally from 25 to 33 $^o/_{oo}$; during rainy weather the water in them is often replaced with strictly fresh water. During dry weather, if the pools are filled with seawater, evaporation occurs until the algae are covered with salt crystals; or if they are filled with freshwater, the algae become bone-dry and brittle after evaporation. During the winter they are subjected to the entire range of air temperatures, sometimes below freezing; during summer the temperature of the upper surface of the rocks becomes so hot that a barefooted person cannot walk upon them. *Calothrix* does not alternately die and recolonize, but tolerates all these conditions. In addition, the *Calothrix* is extensively walked upon by countless sport fisherman, occasionally with regret, as a wet mat of bluegreen algae (students: please note) is at least as slippery as ice.

TOXIC BLUEGREENS

Many bluegreen algae are known to produce one to several toxic compounds that behave like endotoxins and are released only upon death or lysis of the cells. Most of these are freshwater species (Gorham 1960, 1962), but some marine bluegreens are also toxic.

The marine planktonic bluegreen, *Oscillatoria erythraea*, has been known to kill marine animals, including sharks, at least indirectly. Clogging of the gills and suffocation may occur in fish that cannot escape a dense bloom of *O. erythraea*, but there is also circumstantial evidence that toxic compounds develop when the cells of *O. erythraea* undergo lysis.

Ciguatera is a human disease, often fatal, that is caused by eating poisonous marine fish in tropical waters. It is well known in the West Indies and many islands of the tropical Pacific. Randall (1958) has expressed the opinion that the origin of ciguatera toxin is bluegreen algae.

Around the island of St. John, Virgin Islands, the fishes in certain bays are toxic or potentially toxic. In other bays, only a few miles distant, the same kinds of fishes are not toxic. In the bays in which fishes are known to be toxic, not all individuals of a given species are poisonous. The toxicity must come from the environment, through the food. Apparently the toxin is a highly stable compound as it seems to move up the food chain. It must be benthic in origin as plankton-feeding fish are never toxic. If it is an alga, it must be small, as some toxic fishes cannot feed on large algae. Bluegreen algae seem to be the most likely source because some are known to produce toxins and they are abundant in all areas in which ciguatera develops, but the evidence is circumstantial. Most benthic-feeding fishes appear to avoid bluegreen algae but they eat them inadvertently as virtually all other algae have epiphytic bluegreens. Furthermore, fishes eat small invertebrates that may feed extensively on bluegreen algae.

COLLECTION AND PRESERVATION

Thorough collecting of bluegreen algae in the marine environment often requires removal of some of the substrate. In the case of limestone or other types of rocks or seawalls, a geologist's hammer is a handy tool. In the case of wooden structures, a large knife or hatchet, by means of which a thin layer of wood can be removed, is essential.

On intertidal sand beaches protected from heavy wave action, a sample of sand should be obtained to a depth of approximately 2 cm. In the laboratory, the sample should be placed in a screw-capped vial and shaken with several volumes of filtered seawater. Immediately after the sand has settled, the supernate should be poured off into another container, as most of the bluegreens will be in suspension in it. The cells or trichomes can be concentrated by centrifugation or simply allowed to settle, if the suspension is not dense with them. Recognition of bluegreen algal microhabitats and the various niches they occupy will come with field-sampling experience.

Bluegreen algae are readily preserved by simply air drying, with a brief rinse with freshwater to remove the saltwater and prevent formation of crystals. Since material placed in herbaria may remain alive after drying for long periods of time, it is advisable to avoid preservation in formaldehyde or other poisons. If preservation with formaldehyde is necessary to prevent decomposition in transit, use 5 percent U.S.P. formalin in seawater or in water from the place of collection. Five percent formaldehyde is made by combining one part U.S.P. formalin with nineteen parts seawater. In making these dilutions, U.S.P. formalin is taken to be 100 percent, although its actual formaldehyde content is from 33 to 40 percent. Five percent formaldehyde is an optimum concentration for the preservation of all kinds of marine algae. If they are to be kept in it for long periods of time, it is best to add a pinch of "Borax" or sodium borate to keep the pH on the alkaline side. Preserved algae should be kept in the dark as much as possible, as fading of the color is rapid in formaldehyde. Herbarium specimens can be prepared from preserved material. Once dry, the color of the specimens will be preserved indefinitely unless they are in strong light for prolonged periods of time.

MICROSCOPIC EXAMINATION

The preparation of a slide for microscopic examination of bluegreen algae often requires careful separation on the slide of plant masses. This is best done by two dissecting needles before the cover glass is added. Separation from debris or other algae, and its removal from the slide, may be done in the same manner.

If previously dried material is being examined, the addition of a drop of household detergent solution may speed up water absorption and return of the cells to their original condition. The detergent is also effective as an aid in removing granules or flakes of white, polymerized formaldehyde from material stored for some time in liquid preservative.

To improve visibility of cross walls and protoplasmic granules, staining with Gram's iodine is helpful (a saturated solution of iodine in 1 percent potassium iodide). Cross walls and thickened end walls can be seen more readily also if a drop of sulfuric acid-potassium dichromate glass cleaning solution is added briefly. The solution should be tipped off the slide and the algae rinsed with water after a minute or so and before adding the cover glass.

CYANOPHYTA

KEY TO THE GENERA OF THE CYANOPHYTA

1 Plants unicellular or colonial, not forming a true filament (Coccogonales)   2

1 Plants filamentous (Hormogonales)   7

2 Cells not basally attached to a substratum (except by the sheath); cell division producing daughter cells equal in size (Chroococcaceae)   3

2 Some cells of colony or plant basally attached to the substratum and often penetrating it; daughter cells of unequal size; endospores often produced (Chamaesiphonaceae)   *Entophysalis* (p. 60)

3 Cells elongate, ovoid to cylindrical, dividing in a plane perpendicular to the long axis   *Coccochloris* (p. 48)

3 Cells not elongate or only slightly so; ovoid, discoid, or pyriform; division not restricted to a plane perpendicular to the long axis   4

4 Cells disk-shaped, in a single row in the sheath, forming a pseudofilament   *Johannesbaptistia* (p. 51)

4 Cells spherical to slightly elongate or pyriform   5

5 Cells in two-directional rows in the gelatinous matrix
     Agmenellum (p. 52)

5 Cells not in two-directional rows   6

6 Cells radially arranged in the gelatinous matrix, pyriform or heart-shaped   *Gomphosphaeria* (p. 53)

6 Cells irregularly arranged in the matrix, usually spherical or ovoid
     Anacystis (p. 55)

7 Plants without heterocysts (Oscillatoriaceae)   8

7 Plants with heterocysts   13

8 Trichomes spiral, without cross walls   *Spirulina* (p. 65)

8 Trichomes with cross walls (multicellular)   9

9 Numerous granules forming a layer against the cross walls   10

9 Without granules (or at most, two) against the cross walls   11

44

10 End wall of tip cell thickened if mature *Microcoleus* (p. 80)

10 End wall of apical cell always thin *Arthrospira* (p. 74)

11 End wall of apical cell thickened if mature *Oscillatoria* (p. 66)

11 End wall of apical cell always thin 12

12 Only the apical cell tapering *Schizothrix* (p. 70)

12 Trichomes tapering at ends through several to many cells (unless broken) *Porphyrosiphon* (p. 76)

13 Trichomes unbranched (Nostocaceae) 14

13 Trichomes branched 17

14 Trichomes long-tapering and hair-tipped *Calothrix* (p. 84)

14 Trichomes not long-tapering, not hair-tipped 15

15 Trichomes not constricted or only slightly constricted at the nodes, the apex tapering through several cells *Scytonema* (p. 85)

15 Trichomes deeply constricted at the nodes 16

16 Terminal vegetative cells spherical or with rotund end or outer wall when mature *Nostoc* (p. 89)

16 Terminal vegetative cells conical when mature *Anabaina* (p. 88)

17 Trichomes always embedded in limestone; branches of two kinds, cylindrical and tapering to a hairlike tip *Mastigocoleus* (p. 91)

17 Trichomes not penetrating limestone; branches often V-shaped, tapering to a hairlike tip *Brachytrichia* (92)

CLASS MYXOPHYCEAE

Order Coccogonales

The coccoid Myxophyceae are unicellular or colonial. In colonial species, the cells are held together, often with considerable space between them, by a gelatinous polysaccharide secreted by the cells from all wall surfaces. The cells are spherical, oval, or elongate. Reproduction is principally by fragmentation of the colony or the breaking loose from the colony of individual cells. Members of one family, the Chamaesiphonaceae, produce so-called endospores, small vegetative cells that resemble spores.

Johannesbaptistia pellucida, a member of the family Chroococcaceae, forms pseudofilaments that are uniseriate. *Entophysalis deusta* in the Chamaesiphonaceae forms pseudofilamentous rows of cells when it penetrates limestone. The pseudofilaments of these two species differ from true filaments in that the cells become separated from each other soon after division as a result of secretion of polysaccharide material between the cells as well as around their outside walls.

Drouet and Daily (1956) recognized three families of coccoid bluegreens: Chroococcaceae, Chamaesiphonaceae, and Clastidiaceae. Since the latter is restricted to fresh water, only the first two are treated here. They are separated as indicated in the key to the families that follows.

Cell division resulting in daughter cells of equal size, the cells not penetrating the substrate   Chroococcaceae

Cell division usually unequal, the upper daughter cell smaller; cells often penetrating the substrate; "endospores" produced
Chamaesiphonaceae (p. 58)

Family Chroococcaceae

Cells spherical, discoid, or ovoid to oblong, occurring as single cells or arranged in a regular or irregular order within the matrix of coalesced sheath material to form colonies of various sizes and shapes. Colonies sometimes globular and loosely attached or form a layer or cushion attached to the substratum by the sheath material. Cell division produces daughter cells of equal size. Cells do not penetrate the substratum. Hydrolysis of the polysaccharide sheath may occur with release of free single cells.

Drouet and Daily (1956) recognize six genera, all but one of which (*Microcrocis*) occur in brackish or marine environments. Complete synonymy is given for each species.

KEY TO THE GENERA OF THE CHROOCOCCACEAE

1 Cells elongate, ovoid, or cylindrical, dividing in a plane perpendicular to the long axis   *Coccochloris*

1 Cells not elongate or only slightly so; spherical, ovoid, discoid, or pyriform; division in any plane   2

2 Cells disk-shaped, in a single row in the sheath forming a pseudofilament   *Johannesbaptistia* (p. 51)

2 Cells spherical to slightly elongate or pyriform   3

3 Cells in two-directional rows in the gelatinous matrix
                                                Agmenellum (p. 52)

3 Cells not in two-directional rows   4

4 Cells radially arranged in the gelatinous matrix, pyriform or heart-shaped   *Gomphosphaeria* (p. 53)

4 Cells irregularly arranged in the gelatinous matrix, spherical or ovoid
                                                Anacystis (p. 55)

Coccochloris

Cells elongate, ovoid to cylindrical, usually embedded in a gelatinous matrix to form a mass or colony in which the cells are irregularly arranged. Cell division is always in a plane at right angles to the long axis.

KEY TO THE SPECIES OF COCCOCHLORIS

1   Cells 4–8 mm in diameter, to three times as long   *C. stagnina*

1   Cells relatively longer   2

2   Cells 1–3 mm in diameter, to 12 times as long   *C. peniocystis*

2   Cells 2–6 mm in diameter, to 8 times as long   *C. elabens* (p. 50)

Coccochloris stagnina Drouet and Daily.   Colonies bluegreen, olive, yellowish, or brownish; the cells embedded in a gelatinous matrix. Mature cells divide into two truncate-hemispherical or truncate-ovoid daughter cells that become ovoid to elliptical and up to three times as long as broad. Sheaths at first hyaline, often becoming yellowish or brownish with age.

Known along the entire Atlantic and Pacific coasts of North America and from Hawaii and the West Indies, especially in brackish water as an epiphyte or on solid substrata and also in the plankton. Of worldwide distribution.

Tilden 1910, p. 32, pl. 2, fig. 15 (as *Aphanothece stagnina* [Sprengel] A. Braun); Fremy 1938, p. 4 (as *Gloeothece rupestris* [Lyngbye] Bornet); Drouet and Daily 1956, p. 15, fig. 145–163; Humm and Caylor 1957, p. 234; Aziz and Humm 1962, p. 56; Cocke 1967, p. 5, fig. 2–8.

Coccochloris peniocystis Drouet and Daily.   Colonies variously colored (green, bluegreen, pink, violet, rose, red-brown) depending upon the environment. Cells 1–3 mm in diameter and up to 12 diameters long; they are usually curved with the ends rounded or attenuate-conical.

Intertidal and below and in the plankton; although most collections are from freshwater, it is occasionally found in marine habitats along the entire Atlantic and Pacific coasts of North America. Of worldwide distribution.

Kützing 1849–1869, 1:25, pl. 36, fig. 7 (as *Gloeocapsa peniocystis*); Tilden 1910, p. 25, pl. 2, fig. 1–2 (as *Gloeothece linearis* Nägeli); Howe 1920, p. 620 (as *Gloeothece rupestris* [Lyngbye] Bornet); Drouet and Daily 1956, p. 31, fig. 170–172; Aziz and Humm 1962, p. 56; Cocke 1967, p. 7, fig. 11–12.

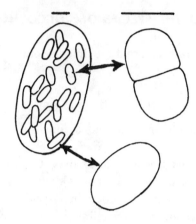

Figure 1. *Coccochloris stagnina.* At left a colony enclosed in a colorless polysaccharide matrix; upper right, a cell in the final stages of division; lower right, a vegetative cell approaching the stage at which division occurs. The lines are 5 microns. From Humm (1979), The Marine Algae of Virginia, with permission of the University Press of Virginia.

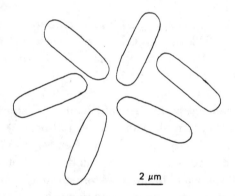

2 μm

Figure 2. *Coccochloris peniocystis.* The line is 2 microns.

Coccochloris elabens Drouet and Daily.   Colonies consist of cells embedded in a hyaline, gelatinous matrix; olive-green, brown, yellow-brown, violet, or rose, depending upon the environment. Cells elliptical to cylindrical, 2–6 μm in diameter and up to 8 diameters long with the ends rounded-truncate.

On moist surfaces and in shallow fresh, brackish, or marine waters and in the plankton. This species is probably much more widely distributed

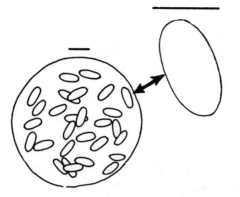

Figure 3. *Coccochloris elabens.* At left a colony enclosed in a spherical, colorless polysaccharide matrix; at right, an enlarged single cell. The lines are 5 microns.

than published records indicate, as it is easily over-looked unless it is abundant.

Kützing 1849–1869, 1:6, pl. 8, fig. 1; Tilden 1910, p. 35, pl. 2, fig. 19 (both as *Microcystis elabens* [Meneghini] Kützing); Drouet and Daily 1956, p. 28, fig. 164–169; Chapman 1961, p. 16, fig. 2; Aziz and Humm 1962, p. 56; Cocke 1967, p. 6, fig. 9–10; Dawes 1974, p. 50, fig. 10.

Johannesbaptistia

Colonies consist of a single row of disk-shaped cells enclosed in a cylindrical gelatinous sheath to form a pseudofilament up to 20 μm in diameter. Cell division occurs only in the plane of the diameter and this results in the uniseriate arrangement in the sheath. There is but one species.

Johannesbaptistia pellucida Taylor and Drouet. Found in shallow, fresh-, brackish, or saltwater, this species has been recorded from New England to the West Indies and Bermuda along the Atlantic coast of North America, and southward to Brazil. From the California coast and Ecuador; probably of worldwide distribution. Taylor (1928) found it in abundance among masses of filamentous algae, especially bluegreens, in pools at Long Key, Florida Keys.

Taylor 1928, p. 48, pl. 1, fig. 23 (as *Nodularia? fusca*); Drouet and Daily 1956, p. 85, fig. 182–184; Cocke 1967, p. 16, fig. 40–41; Dawes 1974, p. 51.

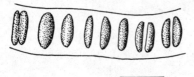

Figure 4. *Johannesbaptistia pellucida.* A short pseudofilament showing several recently divided pairs of cells. The line is 10 microns. From Humm (1979), The Marine Algae of Virginia, with permission of the University Press of Virginia.

10 μm

Agmenellum

This plant occurs as a flat plate, one cell thick, usually microscropic, composed of cells arranged in rows in two directions at right angles to each other. Cell division occurs successively in two planes perpendicular to each other and to the surface of the plate. The sheath is thin and hyaline.

KEY TO THE SPECIES OF AGMENELLUM

Cells 1.0–3.5 μm in diameter, colonies up to 256 cells

A. *quadruplicatum*

Cells 4–10 μm in diameter, colonies larger   A. *thermale*

Agmenellum quadruplicatum Brebisson. Plants bluegreen, olive-green, violet, or rose; flat, rectangular, consisting of a few to 256 cells; cells usually globose, 1.0–3.5 μm in diameter, tightly or loosely arranged in the matrix.

This plant is common on or in wet sand or muddy sand in the intertidal zone of protected, low-energy beaches. Of worldwide distribution.

Tilden 1910, p. 43, pl. 2, fig. 35 (as *Merismopedium glaucum* [Ehrenberg] Nägeli); Drouet and Daily 1956, p. 86, fig. 133; Cocke 1967, p. 17, fig. 42–43; Dawes, 1974, p. 48.

Agmenellum thermale (Kützing) Drouet and Daily. Plants microscopic, sometimes macroscopic, flat, rectangular at first but becoming irregular and up to several centimeters in diameter; bluegreen, olive-green, or some other color as determined by the environment; cells globose, becoming ovoid or cylindrical after division, 4–10 μm in diameter, 4–20 μm long.

In fresh-, brackish, and marine waters, especially in or on sand or muddy sand in protected places. Of worldwide distribution.

Kützing 1843, p. 163 (as *Merismopedia thermalis*); Tilden 1910, p. 42–44, pl. 2, fig. 32, 34, 36 (as *Merismopedium aerugineum, M. novum,* and *M. convolutum*); Fremy 1936, p. 8 (as *Merismopedia convoluta*); Drouet and Daily 1956, p. 89, fig. 134–142; Humm and Darnell 1959, p. 270; Cocke 1967, p. 18, fig. 44–48; Dawes 1974, p. 48, fig. 4; Hamm and Humm 1976, p. 211.

Gomphosphaeria

Plants microscopic, usually spherical; the cells radially arranged in the outer part of the polysaccharide matrix. In addition, each cell may have a more or less distinct sheath of its own. Cells ovoid to pyriform; division in

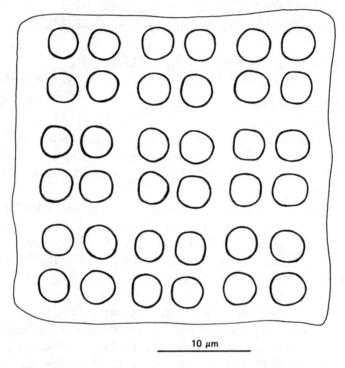

10 μm

Figure 5. *Agmenellum quadruplicatum.* A colony with cells in regular rows. The line is 10 microns.

two planes perpendicular to each other and radial within the matrix. There is only one species.

Gomphosphaeria aponina Kützing.   Plants bluegreen, olive-green, or of some other color as determined by the environment; cells ovoid to pyriform, often heart-shaped during division, 4–15 μm in diameter, radially arranged in the outer layer of the gelatinous matrix which is usually spherical. In older plants there may be dichotomously branched strings from the inner points of the cells to the center of the matrix.

In fresh-, brackish, and marine water in protected places, on or among other algae or in plankton. Of worldwide distribution.

Kützing 1849, p. 233; Tilden 1910, p. 38, pl. 2, fig. 23–28; Taylor 1928, p. 41, pl. 1, fig. 7; Fremy 1936, p. 10, fig. 1; Lindstedt 1943, p. 15, pl. 1, fig. 7; Drouet and Daily 1956, p. 98, fig. 178–180; Chapman 1961, p. 17, fig. 4; Cocke 1967, p. 20, fig. 53–54.

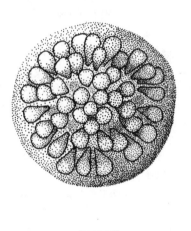

10 μm

Figure 6. *Agmenellum thermale.* A colony in which the cells are close together and in a division stage. The line is 10 microns. From Humm (1979), The Marine Algae of Virginia, with permission of the University Press of Virginia.

10 μm

Figure 7. *Gomphosphaeria aponina.* A colony in which the cells are close together. The line is 10 microns. From Humm (1979), The Marine Algae of Virginia, with permission of the University Press of Virginia.

Anacystis

Cells spherical when mature, dividing successively in three planes perpendicular to each other, often with flattened adjacent faces after division; embedded in a matrix of extracellular polysaccharide forming microscopic to macroscopic colonies, the cells usually irregular in distribution within the matrix but sometimes in a regular cubical arrangement.

KEY TO THE SPECIES OF *ANACYSTIS*

1   Cells 0.5–2.0 μm in diameter   *A. marina*

1   Cells mostly more than 2 μm in diameter   2

2   Cells 2–6 μm in diameter   *A. montana*

2   Cells mostly more than 6 μm in diameter   3

3   Cells 6–12 μm in diameter   *A. aeruginosa* (p. 57)

3   Cells 12–50 μm in diameter, often solitary   *A. dimidiata* (p. 57)

Anacystis marina Drouet and Daily. Plant mass microscopic or visible, bright bluegreen, in spherical, ovoid, or flattened layers; cells hemispherical after division, becoming globose with age, 0.5–2.0 μm in diameter, scattered throughout the gelatinous matrix, often crowded; matrix hyaline, homogeneous, or often diffluent.

Widely distributed as plankton in fresh-, brackish, and sometimes marine waters, often epiphytic on other algae or on solid surfaces. Rarely reported, probably because it is easily overlooked.

Fremy 1936, p. 9 (as *Aphanocapsa marina* Hansgirg); p. 10 (as *Aphanocapsa salina* Woronichin); Drouet and Daily 1956, p. 44, fig. 376–377; Humm and Caylor, 1957, p. 234; Aziz and Humm 1962, p. 56; Cocke 1967, p. 9, fig. 16; Mueller 1976, p. 56.

Anacystis montana (Lightfoot) Drouet and Daily forma *montana* Drouet and Daily. Plants microscopic to macroscopic, variously colored depending upon the environment, planktonic or in a layer or mass upon a substratum. Cells spherical before division, 2–6 μm in diameter, and irregularly distributed or in cubical arrangement within the sheath.

In strata upon solid surfaces and in globular or irregular masses upon other algae. Of worldwide distribution in freshwater and also in brackish water habitats.

Lyngbye 1819, p. 206, pl. 69, fig. B 1–2; Tilden 1910, p. 18, pl. 1, fig. 21 (as *Gloeocapsa conglomerata* Kützing); Taylor 1928, p. 40, pl. 1, fig. 2 (as *Gloeocapsa fuscolutea* [Nägeli] Kirchner); Drouet and Daily 1956, p. 45, fig. 16–90A; Cocke, p. 10, fig. 17–25; Dawes 1974, p. 50, fig. 9.

56

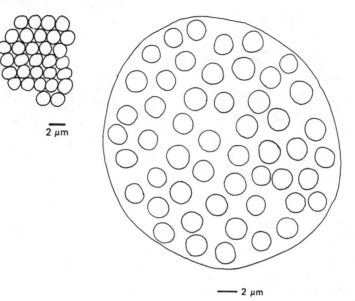

2 µm

— 2 µm

Figure 8. *Anacystis marina.* At left, a layer of cells as seen on the surface of a shell; at right, a colony enclosed in a colorless polysaccharide matrix. The lines are 2 microns. From Humm (1979), The Marine Algae of Virginia, with permission of the University Press of Virginia.

Anacystis aeruginosa Drouet and Daily.   Plants forming microscopic or macroscopic colonies. Cells 6–12 µm in diameter with the adjacent faces flattened after division but becoming spherical before division; irregularly arranged in the colorless gelatinous matrix, cell contents bluegreen in shallow water, often reddish or maroon in deeper water.

Cosmopolitan in shallow marine or brackish water in protected places, usually attached by the sheath to solid surfaces.

Drouet and Daily 1956, p. 76, fig. 108–113; Humm 1979, p. 27, fig. 7.

Anacystis dimidiata (Kützing) Drouet and Daily.   Plants microscopic, one-celled or in colonies of two to four cells (rarely more), bluegreen, with adjacent faces flattened, 12–50 µm in diameter; sheaths hyaline.

Cosmopolitan in shallow fresh-, brackish, and marine waters, usually mixed with other small algae.

Tilden 1910, p. 5 pl. 1, fig. 3; Taylor 1928, p. 39, pl. 1, fig. 4 (both as *Chroococcus turgidus* [Kützing] Nägeli); Humm and Caylor 1957, p.

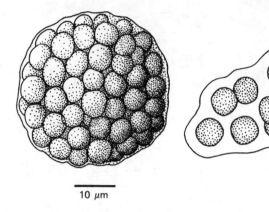

10 µm   10 µm

Figure 9. *Anacystis montana,* forma montana. A colony in which the cells have produced little polysaccharide. The line is 10 microns. From Humm (1979), The Marine Algae of Virginia, with permission of the University Press of Virginia.

Figure 10. *Anacystis aeruginosa.* A small colony. The line is 10 microns. From Humm (1979), The Marine Algae of Virginia, with permission of the University Press of Virginia.

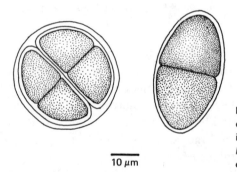

10 µm

Figure 11. *Anacystis dimidiata.* Four-celled and two-celled colonies. The line is 10 microns. From Humm (1979), The Marine Algae of Virginia, with permission of the University Press of Virginia.

232, pl. 1, fig. 2–3; Chapman 1961, p. 16, fig. 3; Cocke 1967, p. 12, fig. 28–37; Dawes 1974, p. 49, fig. 7–8; Humm 1979, p. 27, fig. 8.

Family Chamaesiphonaceae

Plants originate as a single cell attached to solid substrata by the polysaccharide sheath and developing into cushions or strata, the lower cells of which may initiate filamentlike penetration, especially into wood or limestone. Cell division occurs at right angles to the axis of the cell and is unequal; the upper daughter cell is smaller. This cell may emerge from the sheath following its rupture or it may remain in place. Cells of the upper layers may enlarge and divide internally into many small cells, referred to as endospores, which may escape or remain in place. Reproduction also occurs by fragmentation of the colony. There is but one genus, *Entophysalis*.

KEY TO THE SPECIES OF ENTOPHYSALIS

1   On rocks, wood, and shells   *E. deusta*

1   On other algae and living animals   2

2   Cells 1–2 μm in diameter, the plant mass
    yellowish in color   *E. endophytica* (p. 61)

2   Cells mostly over 2 μm in diameter,
    the plant mass bluegreen or some shade of red   *E. conferta* (p. 61)

Entophysalis deusta Drouet and Daily.   Plants producing microscopic or sometimes macroscopic strata or cushions upon various solid surfaces in the intertidal zone or below, usually dark green to blackish in color, but variable. Cells spherical to polyhedral by pressure, 3–6 μm in diameter. Cells that penetrate limestone tend to form irregular filaments 2–15 μm in diameter. Endosporangia spherical to ovoid and up to 30 μm in diameter; the endospores usually 2–5 μm in diameter.

Perkins and Tsentas (1976) found that *E. deusta* (as *Hyella* sp.) penetrated pieces of limestone that they put out at St. Croix, Virgin Islands, at all depths from the intertidal zone to 30 m. It was most abundant in the test limestone samples, however, from the intertidal zone to 15 m.

Of worldwide distribution, at least in temperate and tropical marine waters, and found in these habitats at virtually every locality where microscopic algae are studied. This species had been reported under scores of genera and species, based upon minor variations produced by the environment. Drouet and Daily (1956) give a full synonymy, bibliography, and station list. On rocks, shells, and wood in salt and brackish waters.

Tilden 1910, p. 24, pl. 1, fig. 33 (as *Entophysalis granulosa* Kützing), and pl. 1, fig. 34–36 (as *Chondrocystis schauinslandii* Lemmermann), and p. 51, pl. 3, fig. 9–11 (as *Hyella caespitosa* Bornet and Flahault); Drouet and Daily 1956, p. 103, fig. 185–194; Humm 1952, p. 19; Humm and Caylor 1957, p. 34, fig. 4–5; Chapman 1961, p. 18, fig. 5; Cocke 1967, p. 23, fig. 55–56 a,b.

60

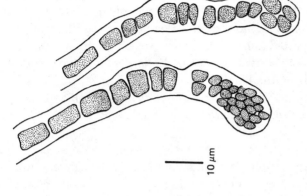

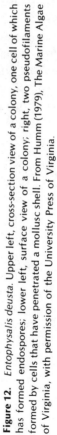

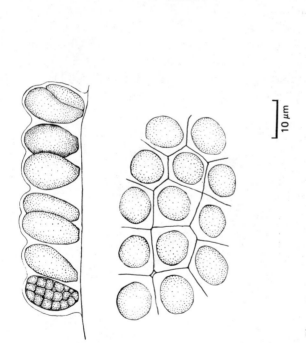

Figure 12. *Entophysalis deusta.* Upper left, cross-section view of a colony, one cell of which has formed endospores; lower left, surface view of a colony; right, two pseudofilaments formed by cells that have penetrated a mollusc shell. From Humm (1979), The Marine Algae of Virginia, with permission of the University Press of Virginia.

Entophysalis endophytica Drouet and Daily.   Plants epiphytic, on the surface and within the polysaccharide outer layer of larger algae; colonies are in the form of small cushions from which penetrating filaments arise. Cells yellow-green, 1–2 μm in diameter, spherical to polyhedral by mutual pressure, with a hyaline sheath. Endosporangia have not been reported. Known from the Pacific Ocean only.

Howe 1914, p. 13 (as *Chlorogloea endophytica*); Setchell and Gardner 1918, p. 434 (as *Chlorogloea lutea*); Drouet and Daily 1956, p. 110, fig. 175.

Entophysalis conferta Drouet and Daily.   Plants microscopic, occasionally macroscopic, epiphytic or endophytic, at first single-celled, then producing a stratum or cushion of various forms. Surface cells mostly 3–6 μm in diameter after division, cells within the host 3–8 μm in diameter. Endosporangia variously shaped, up to 50 μm in diameter, and solitary or in an aggregation of a single layer. Cells typically bluegreen, but may be of other colors. Of worldwide distribution.

Kützing 1845, p. 149; Tilden 1910, p. 50, pl. 3, fig. 7 (as *Xenococcus schousboei* Thuret); p. 52, pl. 3, fig. 13–15 (as *Dermocarpa prasina* [Reinsch] Bornet and Thuret); p. 53 pl. 3, fig. 19–21 (as *D. violacea* Crouan); Drouet and Daily 1956, p. 111, fig. 196–215: Chapman 1956, p. 352, fig. 1, no. 3–5; Humm and Caylor 1957, p. 234, pl. 1, fig. 6–7; Chapman 1961, pl. 18, fig. 6; Cocke 1967, p. 24, fig. 57–58; Dawes 1974, p. 52.

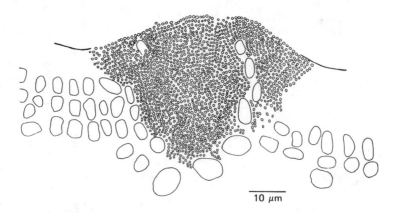

10 μm

Figure 13.   *Entophysalis endophytica*. A colony growing on the surface and among the cells of a host plant (after Howe 1914). The line is 10 microns.

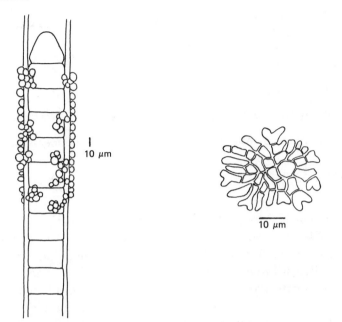

Figure 14. *Entophysalis conferta.* At left, colonies of cells growing upon a filamentous bluegreen host; at right, a disklike colony often found upon larger algae. The lines are 10 microns.

Order Hormogonales

The order Hormogonales includes all the truly filamentous bluegreen algae. The more modern literature divides the order into three families: Oscillatoriaceae, Nostocaceae, and Stigonemataceae. These families are separated on the basis of the presence or absence of heterocysts and the presence or absence of true branching, as shown in the key to the families that follows.

In the older literature, Tilden (1910) and Drouet (1951) recognized five families: Oscillatoriaceae, Nostocaceae, Scytonemaceae, Stigonemataceae, and Rivulariaceae. The Scytonemaceae were characterized as having false branching (branching of the sheath but not of the trichomes); the Rivulariaceae were characterized by a long-tapering and hair-tipped apex and false branching. Desikachary (1959) divides the Hormogonales (as treated here) into two orders: Nostocales, without true branching, and Stigonematales, with true branching. He recognizes six families in the first order (including the Oscillatoriaceae), and seven families in the second.

KEY TO THE FAMILIES OF THE HORMOGONALES

1	Without heterocysts	Oscillatoriaceae
1	Heterocysts produced	2
2	Without true branching	Nostocaceae (p. 83)
2	With true branching	Stigonemataceae (p. 90)

Family Oscillatoriaceae

Plants without heterocysts; the trichomes unbranched without a visible sheath, the polysaccharide material is diffluent to form a common gelatinous matrix, or a firm discrete sheath contains a single trichome or, occasionally, two to many trichomes in a sheath. Trichomes are capable of indeterminate growth in length, although fragmentation occurs by death of a cell or by occasional secretion of polysaccharide material between two cells. The diameter of a trichome may vary as a result of a gradual increase or decrease in cell diameter. The tips of trichomes may taper through a few to many cells or they may be cylindrical; the tip cell may be rounded, truncate, or conical; the outer membrane of the tip cell may be consistently thin or distinctly thickened when the cell is mature or old, one of the characters used by Drouet (1968) to separate genera. In cell division, formation of the cross wall begins by constriction of the outer wall and progresses centripetally toward the center. The cell contents may appear homogeneous or granular, with granules larger or more numerous when nutrients are abundant and conditions favorable for photosynthesis. Some species accumulate many granules against the cross walls, another character used by Drouet (1968) to separate genera.

Probably the most difficult characteristic used to separate the above genera is the condition of the end wall of terminal cells. Only older, mature terminal cells will exhibit thickened end walls in the genera *Microcoleus* and *Oscillatoria* as these take some time to develop after fragmentation of the trichome. It is often necessary to examine many tip cells before a decision can be made.

A similar problem may occur in deciding whether or not the trichomes

64

taper through several terminal cells. Recently broken trichomes with this potential will not show it.

It is unwise to attempt to identify a single trichome or even a small number as they may not possess the important characteristics that they have the potential of developing.

In the older literature the term "calyptra" has been used to imply something more than a thickened end wall. Furthermore, it has been preempted in botany in the taxonomy of mosses.

Spirulina

Trichomes typically spiral, rarely simply curved or straight, apparently without cross walls; the ends rounded and the end walls not thickened; the extracellular polysaccharide usually invisible, often present as a soft, diffluent mass. Trichomes motile and rotating. Reproduction by median division only. Drouet (1968) recognizes only one species.

Spirulina subsalsa Oersted.   Plant mass or trichomes exhibiting a wide range of colors depending upon light conditions, although often bright bluegreen in strong light and shallow water and red or maroon in light that has filtered through one or more meters of water. Trichomes, without visible (by light microscope) cross walls, usually distinctly spiral and 0.4–4.0 μm in diameter (usually 1–2 μm in marine forms). Trichome length is apparently determined by the environment. Trichomes increase

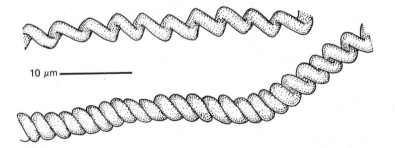

Figure 15. *Spirulina subsalsa.* Two filaments (single cells, actually) showing different degrees of spiraling. The line is 10 microns. From Humm (1979), The Marine Algae of Virginia, with permission of the University Press of Virginia.

in number by fragmentation. As a result of constriction of the cell walls toward the center of the cell, new walls are formed that become the end walls when the trichome breaks.

Common in shallow seawater along both coasts of North America and widely distributed in freshwater. Of worldwide distribution. Drouet (1968) expresses the opinion that the higher the salinity the tighter the spiral. Loosely spiralled forms and straight trichomes may be confined to freshwater or water of low salinity.

Oersted 1842, p. 17, pl. 7, fig. 4; Gomont 1892, p. 253, pl. 7, fig. 32; Tilden 1910, p. 89, pl. 4, fig. 49; Taylor 1928, p. 46, pl. 1, fig. 12; Humm and Caylor 1957, p. 235, pl. 1, fig. 11; Chapman 1961, p. 21, fig. 1; Drouet 1968, p. 333, fig. 1–7; Dawes 1974, p. 58, fig. 20; Humm 1979, p. 36, fig. 14.

Oscillatoria

Trichomes without heterocysts and without granules on the cross walls (or at most, two); the end wall of the terminal cell becoming thick with age, the tips of the trichomes often attenuated, the terminal cell hemispherical to somewhat conical with a rounded end. Trichomes straight, curved, or spiral, producing an extracellular polysaccharide that may be invisible, soft and diffluent, or forming a firm sheath.

KEY TO THE SPECIES OF OSCILLATORIA

1   Strictly planktonic   *O. erythraea*

1   Not strictly planktonic   2

2   End wall of terminal cell only slightly convex; cells usually shorter than wide   *O. lutea* (p. 68)

2   End wall of terminal cell hemispherical to short-conical; cells usually isodoametric   *O. submembranacea* (p. 69)

Oscillatoria erythraea (Ehrenberg) Kützing. Plants strictly plank-tonic and marine in the form of bundles of trichomes held together by diffluent extracellular polysaccharide, the bundles breaking apart as their size reaches the limit of the adhesive properties of the polysac-charide, often 1 mm in diameter and one to several millimeters long and visible to the unaided eye. Trichomes 3–20 μm in diameter, the cells 2–27 μm long, often with large gas vacuoles, and actively moving back and forth within the bundles. Terminal cells cylindrical with rounded end, conical with a truncate end, or swollen and bulbous, the end wall becom-ing thickened with age.

This species annually forms vast areas of discolored water or "red tide" in the open sea during the warmer months of the year, especially when the waters are lowest in nitrogen salts, in the Sargasso Sea, the Caribbean Sea, the Gulf of Mexico, among the Hawaiian islands, Japanese waters, and other tropical or subtropical seas. The macroscopically visible scalelike colonies have been called "sea sawdust" and color the water, when abundant, yellowish, amber, or red. They are of major importance in basic productivity of these waters and have been shown to fix nitrogen. Toxic compounds that may cause fish kill apparently are produced when the cells lyse, although some hold the opinion that fish kill is a result of gill clogging.

Gomont 1892, p. 197, pl. 6, fig. 2–4; Tilden 1910, p. 84, pl. 4, fig. 41–42; Collins and Hervey 1917, p. 20; Howe 1920, p. 621; Taylor 1928, p. 47, pl. 1, fig. 15; Fremy 1934, p. 114, pl. 30, fig. 3; Humm and Caylor 1957, p. 236, pl. 2, fig. 8; all as *Trichodesmium thiebautii* Gomont; Drouet 1968, p. 212, fig. 130–131; Dawes 1974, p. 54, fig. 15.

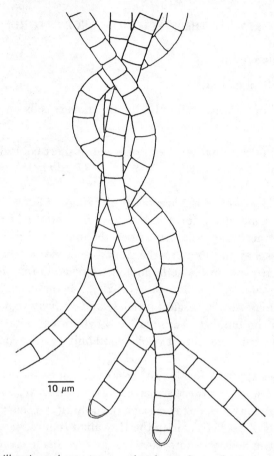

10 µm

Figure 16. *Oscillatoria erythraea.* A group of trichomes from a fasciculate colony found in open-sea plankton. The line is 10 microns.

Oscillatoria lutea C. Agardh.   Trichomes 2.5–10.0 µm in diameter; the cells 1–7 µm long, usually considerably shorter than the diameter but sometimes as long as broad, widely variable in color, sometimes with slight constrictions at the nodes, and without granules on the cross walls. Apical cell somewhat conical with a truncate end and developing a thickened end wall with age. Trichomes cylindrical at the tip or tapering slightly through the last few cells; the extracellular polysaccharide hyaline, invisible, forming a diffluent layer or a distinct sheath with one to many trichomes; sheaths sometimes branched.

Widely distributed in salt- and freshwater and on damp soil in

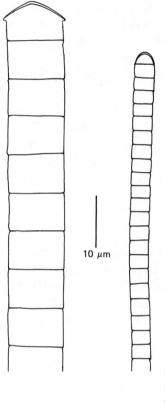

10 μm

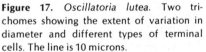

10 μm

Figure 17. *Oscillatoria lutea.* Two trichomes showing the extent of variation in diameter and different types of terminal cells. The line is 10 microns.

temperate and tropical regions. There are many marine records along both coasts of North America, including the Gulf of Mexico, the West Indies, and Hawaii. Of worldwide distribution.

Gomont 1892, p. 141, pl. 23, fig. 12–13; Tilden 1910, p. 114, pl. 5, fig. 30–31; Hoyt 1917-18, p. 413; Howe 1924, p. 358; Fremy 1934, p. 109, pl. 28, fig. 4; 1936, p. 25; 1938, p. 30; Newton 1931, p. 24; Lindstedt 1943, p. 86, pl. 11, fig. 1–2; Desikachary 1959, p. 310, pl. 52, fig. 9; Cocke 1967, p. 59, fig. 134 (all as *Lyngbya lutea* Gomont); Drouet 1968, p. 185, fig. 52; Baker and Bold 1970, p. 42, fig. 95–96; Dawes 1974, p. 55.

Oscillatoria submembranacea Ardissone and Strafforella. Trichomes 2.5–9.0 μm in diameter; the cells 3–11 μm long, isodiametric or a little shorter or longer than the diameter. Their color is variable, although usually bluegreen. The apical cell is cylindrical with rounded end or somewhat conical-truncate; the end wall becomes thickened with maturity. The sheath is apparently absent or diffluent, or firm and distinct, sometimes branched, with one to many trichomes.

Widely distributed in salt- and freshwater in temperate and tropical regions, occasional in cold waters.

Gomont 1892, p. 129, pl. 2, fig. 5; Tilden 1910, p. 129, pl. 5, fig. 48; Humm and Caylor 1957, p. 238, pl. 2, fig. 11; Cocke 1967, p. 65, fig. 148 (all as *Symploca atlantica* Gomont); Gomont 1892, p. 180, pl. 5, fig. 13; Tilden 1910, p. 104, pl. 5, fig. 6 (all as *Phormidium submembranaceum* [Ardissone and Strafforella] Gomont); Drouet 1968, p. 203, fig. 62–64; Humm 1976, p. 74.

Schizothrix

Trichomes without heterocysts and without granules on the cross walls (or at most, two), sometimes varying in diameter in a single trichome, and not tapering at the tips. Apical cells conical to hemispherical; the end wall not thickened. Filaments straight, curved, or sometimes spiral. External polysaccharides invisible, evident as a soft diffluent mass, or forming a firm distinct sheath.

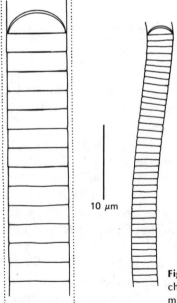

10 μm

Figure 18. *Oscillatoria submembranacea.* Two trichomes showing variation in cell length. The line is 10 microns.

KEY TO THE MARINE SPECIES OF SCHIZOTHRIX

1   Mature terminal cell hemispherical   2

1   Mature terminal cell tapered, conical   3

2   Trichomes 1.0–3.5 µm in diameter   *S. calcicola*

2   Trichomes 4–60 µm in diameter   *S. mexicana* (p. 73)

3   Mature terminal cell long-acuminate   *S. tenerrima* (p. 73)

3   Mature terminal cell short-tapered, conical, with rounded tip
                                                                       S. arenaria (p. 73)

Schizothrix calcicola (C. Agardh) Gomont.   Trichomes 0.2–3.5 µm in diameter, the cells 0.2–6.0 µm long, the end wall of mature terminal cells rounded to hemispherical, not thickened. Cross walls sometimes with one or two granules on each side; the nodes may or may not be constricted. Extracellular polysaccharides invisible, or, if visible, soft and diffluent (as in the old genus *Phormidium*), or forming a distinct sheath (as in the old genus *Lyngbya*), or forming a firm, cartilaginous mass in which the trichomes are embedded.

  S. calcicola is widely distributed and abundant in virtually all marine algal habitats as well as on land and in freshwater. It survives long periods of dessication. It bores into limestone and is abundant in oyster and clam shells, calcareous worm tubes, and bryozoan tests, penetrating as deeply as light permits. It produces a greenish layer in the limestone of both living and dead stony corals and may contribute organic matter to the coral polyp.

  Perkins and Tsentas (1976) found *S. calcicola* (as *Plectonema terebrans*) to be the second most common bluegreen boring into limestone at St. Croix, Virgin Islands. It was equally common from the intertidal zone to 30 m depth, the range in which they planted skeletal carbonate pieces to study the algae that bore into limestone.

  It is abundant as an epiphyte of larger marine algae and seagrasses and is common on the surface of invertebrate animals as well as nonliving substrata. It is one of the most common species of bluegreen algae in the sea, occurring in virtually every habitat in which marine algae are found. Trichomes are often present among the marine plankton of inshore waters, especially after windy weather.

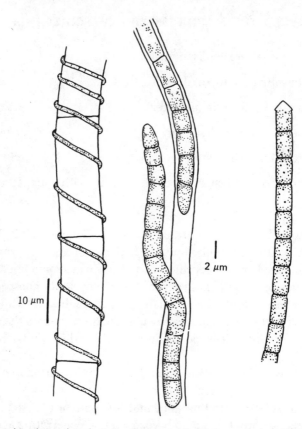

Figure 19. *Schizothrix calcicola.* Left, a filament spiraling around a branch of a green alga, *Cladophora.* The line is 10 microns. Center, a filament showing false branching; right, a trichome with a conical tip cell. The line is 2 microns. From Humm (1979), The Marine Algae of Virginia, with permission of the University Press of Virginia.

Drouet (1963) discussed the influence of the environment upon *Schizothrix calcicola* and recognized its many ecophenes. In the past, bluegreen systematists placed this species in scores of genera and hundreds of species based upon trivial variations in morphology that seem to be a product of the environment. Some of these older species are treated in the appendix.

Drouet 1968, p. 27, fig. 8–19; Baker and Bold 1970, p. 22, fig. 2–4; Dawes 1974, p. 57, fig. 19.

Schizothrix mexicana Gomont.   Trichomes 4–65 μm in diameter; the cells 2–10 μm long, usually shorter than the diameter or more or less isodiametric, sometimes a little longer, the terminal cell with rounded end and a thin end wall. Trichomes not attenuate at the tips, with or without constrictions at the nodes. Extracellular polysaccharide thin and invisible, or obvious and soft-diffluent, or forming a distinct sheath containing one to many trichomes; the sheath sometimes branched.

Of worldwide distribution in salt- and freshwater of temperate and tropical regions. Some marine forms produce trichomes of large diameter.

Gomont 1892, p. 124, pl. 2, fig. 20; Tilden 1910, p. 117, pl. 5, fig. 36 (as *Lyngbya gracilis*); p. 118, pl. 5, fig. 37 (as *L. sordida*); Drouet 1968, p. 87, fig. 20–22.

Schizothrix tenerrima (Gomont) Drouet. Trichomes 1–6 μm in diameter, usually constricted at the nodes; the cells 3–12 μm long, the cross walls not granulated. Terminal cells becoming long-conical with maturity and often with a hairlike tip; the outer membrane thin. The extracellular polysaccharide may be thin and invisible or it may be persistent and diffluent, or it may produce a distinct tubular sheath with from one to, more often, many trichomes within. The sheath may be branched.

Schizothrix tenerrima is typically a plant of the intertidal zone on bottom sediments, rocks, or epiphytic. In the absence of a distinct sheath, the mature, attenuate terminal cells are of major importance in distinguishing this species from *S. calcicola* and *S. arenaria*.

Gomont 1892, p. 355, pl. 14, fig. 9–11 (as *Microcoleus tenerrimus*); Drouet 1968, p. 135, fig. 42–45.

Schizothrix arenaria (Berkeley) Gomont. Trichomes 1–6 μm in diameter; the cells isodiametric or longer than wide, 2–10 μm long; the tip cells distinctly conical and with a thin membrane and a broadly rounded end. Cross walls slightly to distinctly constricted, without granules or sometimes with a single granule near each wall. Extracellular polysaccharides may be diffluent and invisible or there may be a distinct sheath containing one to many trichomes. Branching of the sheath may occur.

Of worldwide distribution in both salt- and freshwater, and often mixed with *Microcoleus lyngbyaceus* in salt marshes and on tidal flats.

Gomont 1892, p. 312, pl. 8, fig. 11–12; Drouet 1968, p. 109, fig. 28–34; Baker and Bold 1970, p. 30, fig. 49–53; Dawes 1974, p. 56, fig. 17–18.

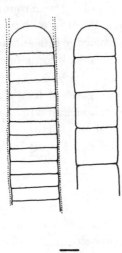

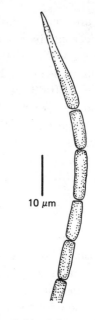

Figure 20. *Schizothrix mexicana.* Tips of two trichomes showing variation in cell length. The line is 10 microns.

Figure 21. *Schizothrix tenerrima.* Tip of a trichome showing a typical elongated and tapering end cell. The line is 10 microns.

Arthrospira

Trichomes with a dense layer of granules on both the cross walls and the lateral walls; the tips often tapering through several cells the terminal cell at first with rounded end but conical when mature; the end wall not thickened. Filaments torulose, curved to spiral; the extracellular polysaccharide thin and invisible, diffluent and evident, or forming a distinct cylindrical sheath.

Drouet (1968) recognized two species, one of which (*A. jenneri* [Hassall] Stizenberger) is found almost exclusively in freshwater.

Arthrospira neapolitana (Kützing) Drouet.   Trichomes 2–10 μm in diameter; the cells shorter than the diameter, 1.5–4.0 μm long, except for the mature apical cells that may be to six times as long as their basal diameter, acute-conical, and sometimes curved. Trichomes may be con-

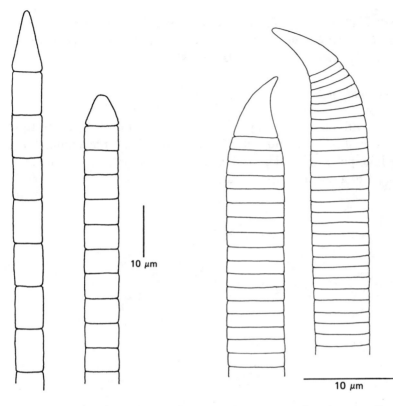

Figure 22. *Schizothrix arenaria.* Two trichomes showing variation in cell length and
degree of elongation and tapering of the tip
cell. The line is 10 microns. From Humm
(1979), The Marine Algae of Virginia, with
permission of the University Press of Virginia.

Figure 23. *Arthrospira neapolitana.* The
line is 10 microns.

stricted at the nodes in some segments, not constricted in others. Extracellular polysaccharides thin and invisible, or diffluent, or forming a
distinct sheath, sometimes branched, with one to many trichomes.

In fresh- and brackish water of temperate and tropical regions.

Drouet (1968) used the name *A. brevis* (Kützing) Drouet for this species
but later (1969) pointed out that *A. neapolitana* is the valid name.

Kützing 1843, p. 186 (as *Oscillaria neapolitana*); Gomont 1892, p. 229,
pl. 7, fig. 14–15; Tilden 1910, p. 79, pl. 4, fig. 32; Newton 1931, p. 18,

Lindstedt 1943, p. 63, pl. 7, fig. 3; Chapman 1961, p. 24, fig. 13 (all as *Oscillatoria brevis* Kützing); Drouet 1968, p. 219, fig. 86–88 (as *Arthrospira brevis* [Kützing] Drouet); 1969, p. 339; Humm 1976, p. 74.

Porphyrosiphon

Trichomes without granules on the cross walls; attenuated at the tips through several cells, the end walls of mature apical cells remaining, thin. Extracellular polysaccharides varying in condition from thin and invisible to diffluent to a distinct, firm sheath.

KEY TO THE SPECIES OF PORPHYROSIPHON

1   Trichomes 1–4 μm in diameter   *P. miniatus*

1   Trichomes larger   2

2   Trichomes 6–12 μm in diameter, cells less than one-third the diameter in length   *P. kurzii*

2   Trichomes 3–40 μm in diameter, cells one-third as long as the diameter or longer   *P. notarisii* (p. 79)

Porphyrosiphon miniatus (Hauck) Drouet. Trichomes 1–4 μm in diameter; the cells quadrate or a little longer than the diameter, 1.5–7.0 μm long, occasionally constricted at the nodes, although usually not. Trichomes taper at the tips through several cells, the mature tip cell conical, often with swollen or bulbous tip, and the end wall not thickened. Sheath thin, invisible, or in the form of a diffluent, hyaline layer.

This species is found only in subtropical and tropical marine waters that are clean and of high salinity. It is known along the Atlantic coast of North America from North Carolina to the Caribbean Sea and in the Gulf of Mexico. It occurs in shallow water, forming yellow-brown streamers or delicate groups of filaments on stones, shells, worm tubes, seagrass leaves, and larger algae. The type specimen came from near Trieste, Italy, in the Mediterranean. It is also known from Biarritz, France, and Japan. It is probably in all tropical seas.

Hauck 1878, p. 80, pl. 1, fig. 16–17 (as *Spirulina miniata*); Gomont 1892, p. 248 (as *Arthrospira miniata* Gomont); Williams 1948, p. 684 (as *Oscillatoria williamsii* Drouet); Umezaki 1961, p. 64 (as *Arthrospira miniata, forma acutissima* Umezaki); Humm 1963, p. 517; 1964, p. 308 (as *A. miniata*); Drouet 1968, p. 172, fig. 81–83.

Porphyrosiphon kurzii (Zeller) Drouet. Trichomes 6–12 μm in diameter; the cells 2–4 μm long (to one-third the diameter), usually with some constriction at the nodes; ends of trichomes taper through several cells, the mature apical cell conical with rounded tip and thin end wall. Extracellular polysaccharides thin and invisible, or diffluent into a homogeneous mass or layer, or in the form of a distinct cylindrical sheath that may be branched.

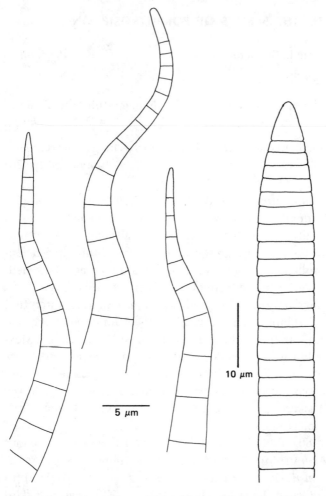

10 µm

5 µm

Figure 24. *Porphyrosiphon miniatus.* Three sharply tapered trichomes. The line is 5 microns.

Figure 25. *Porphyrosiphon kurzii.* The line is 10 microns.

Known from the intertidal zone and below in warm-temperate and tropical seas. Recorded from North Carolina to the West Indies and Gulf of Mexico on the Atlantic coast of North America, and from southern California southward on the Pacific coast. This species is apparently strictly marine.

Some of the older genera in which this plant has been placed in the past (*Sirocoleum, Hydrocoleum*) were defined as having false branching in recognition of the frequency of this condition.

Gomont 1892, p. 349, pl. 14, fig. 3–4 (as *Sirocoleum kurzii*); Gardner 1932, p. 289 (as *Hydrocoleum borgesenii*); Drouet 1968, p. 162, fig. 76.

Porphyrosiphon notarisii (Meneghini) Kützing.   Trichomes 3–40 µm in diameter; the cells 3–15 µm long, shorter or longer than the diameter but more often shorter, usually with constrictions at the nodes. Tips of trichomes when mature taper through several cells, the apical cell slightly to acutely conical, the end broadly rounded to sharply pointed; the end wall not thickened. Extracellular polysaccharide a thin invisible layer, or diffluent into a soft gelatinous mass, or forming a firm distinct sheath, sometimes branched, containing one to many trichomes.

Of worldwide distribution, especially in temperate and tropical regions, in fresh- and saltwater, in hot springs, and on damp soil, rocks, and wood. In the marine environment it occurs in the intertidal zone and below, and has been found essentially the entire length of both Atlantic and Pacific coasts of North America, among the West Indies, and at Hawaii.

Kützing 1863, p. 7; Gomont 1892, p. 217 pl. 6, fig. 10; Tilden 1910, p. 69, pl. 4, fig. 12 (all as *Oscillatoria nigro-viridis* Thwaites); Drouet 1968, p. 143, fig. 65–75; Baker and Bold 1970, p. 40, fig. 89–91; Dawes 1974, p. 56, fig. 16; Ballantine and Humm 1975, p. 155.

Microcoleus

Trichomes usually taper through severa cells; the terminal cell somewhat conical with a rounded or truncate end, the end wall distinctly thickened

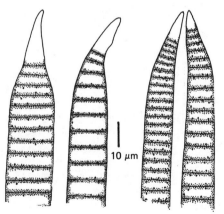

Figure 26. *Porphyrosiphon notarisii.* Four trichomes with well-developed tips. The line is 10 microns.

when mature. End walls of the cells with a conspicuous layer of granules. Extracellular polysaccharide may be thin and invisible, soft and diffluent, or forming firm sheaths containing one to many trichomes.

Drouet (1968) recognized three species, one of which, *M. irriguus* (Kützing) Drouet, is found only in freshwater and is not included here.

KEY TO THE SPECIES OF MICROCOLEUS

Granules lining end walls only; trichomes 2.5–9.0 μm in diameter

M. vaginatus

Granules lining both end and lateral walls; trichomes 3.5–80 μm in diameter

M. lyngbyaceus

Microcoleus vaginatus (Vaucher) Gomont.   Trichomes 2.5–9.0 μm in diameter; the cells isodiametric or shorter or longer than the diameter, 1–10 microns long, rarely constricted at the nodes; the cross walls with a layer of granules on each side. Apical cell when mature usually conical with rounded or truncate end, the end wall distinctly thickened. Trichomes usually somewhat tapered at the ends through several cells. Sheaths thin and invisible or fusing to form a diffluent mass or present as a distinct, firm tube that is sometimes branched and with usually one, sometimes many, trichomes.

Of cosmopolitan distribution in fresh- and somewhat brackish water, although not found growing in full-salinity seawater. It tolerates long periods of dessication.

Gomont 1892, p. 355, pl. 14, fig. 12; Drouet 1962, fig. 1–27; 1968, p. 226, fig. 88–99; Baker and Bold 1970, p. 33, fig. 54–57; Dawes 1974, p. 54.

Microcoleus lyngbyaceus (Kützing) Crouan.   Trichomes 3.5–80.0 μm in diameter, tapering briefly or over considerable length at the tips; usually constricted at the nodes, the transverse and lateral walls lined with granules. Cells usually shorter than their length, from about 1/15 up to (but rarely) as long as the diameter, 1.5–8.0 μm long; the apical cell short-conical with a rounded or truncate end, and the end wall conspicuously thickened. Extracellular polysaccharides thin and invisible, diffluent into a mass or layer, or forming a distinct and firm sheath that may be branched and contain one to many trichomes.

This species is cosmopolitan in fresh-, brackish, and seawater and in a great variety of habitats. In brackish water it may develop gas vacuoles and produce a water bloom. In marine habitats it may produce conspicuous coatings or layers in the intertidal zone, one form exhibiting remarkable resistance to the toxic compounds of creosoted pilings. It may form extensive mats in seagrass beds, segments of which often float as a result of en-

81

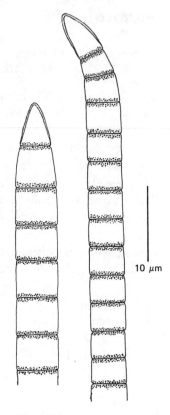

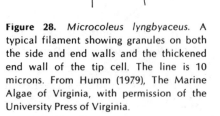

Figure 27. *Microcoleus vaginatus.* The line is 10 microns.

Figure 28. *Microcoleus lyngbyaceus.* A typical filament showing granules on both the side and end walls and the thickened end wall of the tip cell. The line is 10 microns. From Humm (1979), The Marine Algae of Virginia, with permission of the University Press of Virginia.

trapment of oxygen bubbles. These flat mats may enter the Gulf Stream from the West Indies or the Florida Keys and drift northward to Cape Cod. They often become entangled in pelagic *Sargassum* species.

This species is probably the most abundant and of the greatest biomass of any member of the Oscillatoriaceae in the sea with the possible exception of the planktonic *Oscillatoria erythraea.*

Kützing 1843, p. 185 (as *Oscillatoria lyngbyacea*); Crouan 1867, p. 114; Tilden 1910, pp. 118, 119, 123 (and others) as *Lyngbya semiplena* (C.

Agardh) J. Agardh, *L. confervoides* C. Agardh, *L majuscula* (and others), pl. 5, fig. 38, 39, 42; Drouet 1968, p. 290, fig. 101–129; Baker and Bold 1970, p. 37, fig. 81–83.

Family Nostocaceae

Trichomes free or in colonies as determined by the presence or absence of sheaths or a gelatinous matrix; uniseriate, unbranched, the heterocysts terminal and intercalary, spores present or absent. The relative abundance of both heterocysts and spores is strongly influenced by the environment.

In his *Revision of the Nostocaceae with Cylindrical Trichomes*, Drouet (1973) relegated to synonymy scores of genera that were based upon sheath characteristics and other minor differences that vary with environmental conditions. In this group, he recognized only two genera with species that occur in the sea, *Calothrix* and *Scytonema*. A third genus, *Raphidiopsis*, is found only in freshwater.

In his *Revision of the Nostocaceae with Constricted Trichomes*, Drouet (1978) recognizes only two genera, *Anabaina* and *Nostoc*, each of which has one species that is commonly found in marine habitats.

KEY TO THE MARINE GENERA OF THE NOSTOCACEAE

1   Trichomes long-tapering, hair-tipped   *Calothrix*

1   Trichomes not long-tapering or hair-tipped   2

2   Trichomes slightly or not constricted at the nodes   *Scytonema* (p. 85)

2   Trichomes much constricted at the nodes   3

3   Mature terminal vegetative cell becoming conical   *Anabaina* (p. 88)

3   Mature terminal vegetative cell with rounded end   *Nostoc* (p. 89)

Calothrix

Trichomes with basal heterocysts and with or without intercalary heterocysts; the mature or original tips long, tapering, and ending in a hairlike row of slender, colorless cells. In short trichomes, the taper may extend from base to apex. Trichomes unbranched, uniseriate, and the nodes not constricted or only slightly so. Extracellular polysaccharides may fuse to form a gelatinous mass in which the trichomes are embedded, or distinct sheaths may be present that often show false branching and containing one to a few trichomes.

Drouet (1973) recognizes only two species, one of which, *C. crustacea*, is marine, and the other, *C. parietina*, inhabits freshwater.

Calothrix crustacea Schousboe and Thuret.   This species is cosmopolitan in the sea and in brackish water, and grows from the intertidal or splash zone to considerable depths on a great variety of substrata, including larger algae and animals, especially invertebrates. High in the intertidal zone it forms a black band that often resembles the mark of an oil spill, especially in harbors and estuaries. In salt marshes it produces a band on the stems of the salt marsh plants *Spartina* and *Juncus*; in mangrove swamps it occurs as an upper band on the prop roots of red mangroves (*Rhizophora*) and on the aerial roots of black mangroves (*Avicennia*). Along coastal areas of high salinity, it may form small gelatinous spheres (the old genus *Rivularia*). On the pelagic species of *Sargassum* in the Sargasso Sea and the Gulf of Mexico it produces small black patches known as a "tar spot" (the old genus *Dichothrix*), and also on plants of *Sargassum* that originally grew attached but broke loose, drifting out to sea. It can be found on nearly all of the larger red, brown, and green algae on the older parts of the plants in shallow water to at least 30 meters deep and probably much deeper.

Calothrix crustacea, in all of its many forms found in the sea, is probably second only to *Oscillatoria erythraea* (*Trichodesmium*) in importance as a nitrogen-fixer. As a "tar spot" on pelagic *Sargassum* plants in the extremely nutrient-poor surface waters of the Sargasso Sea, it may be one of the reasons that *Sargassum* plants can obtain sufficient fixed nitrogen to permit them to become so abundant. It may have a beneficial effect on all of the larger benthic algae on which it is such a familiar and abundant epiphyte.

Perkins and Tsentas (1976) found that *C. crustacea* (*Calothrix* sp.) bores into skeletal limestone, an ability that apparently had not been recognized previously with this species. They planted limestone pieces from the intertidal zone to a depth of 30 m and found *C. crustacea* penetrating them at all depths but in greatest abundance from 5 to 15 m.

Bornet and Flahault 1886, p. 359; Collins and Hervey 1917, p. 27; Taylor 1928, p. 51, pl. 2, fig. 10; Newton 1931, p. 34, fig. 23; Drouet 1942, p. 69; Blomquist and Humm 1946, p. 3, pl. 1, fig. 6; Fan 1956, p. 172, fig. 5; Williams 1948, p. 684; Almodover and Blomquist 1959, p. 167; Chapman 1961, p. 43, fig. 45; Cocke 1967, p. 156, fig. 304; Drouet 1973, p. 180, fig. 78; Dawes 1974, p. 60, fig. 22–23.

Scytonema

Trichomes cylindrical, straight, curved, or spiral; the ends attenuate through several cells, but not long-tapering or hair-tipped, somewhat constricted at the nodes; sheaths typically much branched, the branches double or single; heterocysts terminal or intercalary, spherical, ovate, to cylindrical; spores ovate to cylindrical; sheaths usually firm and distinct in plants from marine habitats, but the extracellular polysaccharide may be soft and diffluent or absent.

Scytonema hofmannii C. Agardh.   Trichomes 3–30 μm in diameter; the cells isodiametric or longer or shorter than the diameter, 3–30 μm long, and more or less constricted at the nodes. The trichomes may be somewhat swollen at the tips and there may be variations in diameter along a given trichome. Spores cylindrical, the walls becoming yellow or brownish with maturity. Sheaths usually distinct, often laminated with age; false branching common, often originating between two heterocysts, single or in pairs.

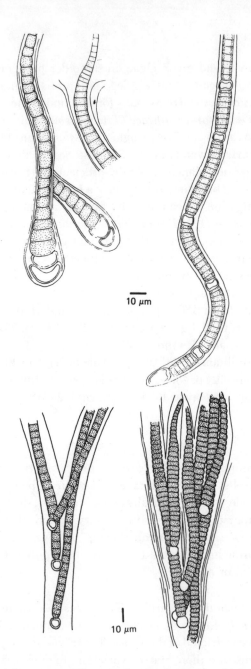

10 μm

10 μm

Figure 29. *Calothrix crustacea.* Filaments showing the basal and intercalary heterocysts, hair tips, false branching, and the occurrence of trichomes in dense fascicles. The line is 10 microns. From Humm (1979), The Marine Algae of Virginia, with permission of the University Press of Virginia.

86

Plant mass often widely expanded and high intertidal in marine habitats or on muddy flats in protected areas; common under mangroves in the tropics. In some areas, such as the Florida Keys and the West Indies, *S. hofmannii* may replace *Calothrix* to form a black zone on rocks high in the intertidal zone.

Perkins and Tsentas (1976) reported *S. hofmannii* (as *Scytonema* sp.) boring into limestone, a propensity not previously reported for this species. It was found in limestone pieces that they placed at various depths around St. Croix, Virgin Islands, from the intertidal zone to 22 meters. It was not found in samples planted from 23 to 30 m.

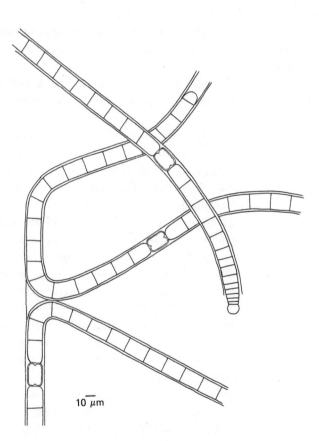

10 μm

Figure 30. *Scytonema hofmannii.* Filaments showing a pair of false branches and intercalary heterocysts. The line is 10 microns.

Of cosmopolitan distribution in fresh- and saltwater and on damp soil.
C. Agardh 1817, p. 117; Bornet and Flahault 1887, p. 97; Tilden 1910, p. 216, pl. 12, fig. 4; Drouet 1973, p. 21, fig. 1–14; Humm and Hamm 1976, p. 43; Hamm and Humm 1976, p. 212.

Anabaina

Trichomes much constricted at the cross walls, moniliform to torulose, uniseriate, straight or curved. Extracellular polysaccharides diffluent, soft and indistinct, or firm-gelatinous and forming a distinct sheath or matrix; cells hemispherical or spherical, the terminal vegetative cells becoming acute- or obtuse-conical with age or maturity; heterocysts terminal or intercalary, spherical, ovoid, or conical; spores usually somewhat larger than the vegetative cells, single, or in a series between heterocysts, spherical, ovoid, or cylindrical with rounded ends.

Drouet (1978) recognizes only two species of *Anabaina*, *A. licheniformis* and *A. oscillarioides*. Only the latter species commonly occurs in the sea.

Anabaina oscillarioides Bory.   Trichomes 2–12 μm in diameter, of indeterminate length; fragmenting by death of a cell, by maturing of a spore, or by continued constriction of a cross wall, blue-green, yellow-green, olive, brown, red, or violet, much constricted at the nodes; plants, when in a visible aggregation, usually forming a greenish-black thin layer; cells spherical or a little shorter or longer than the diameter, 2–10 μm long, heterocysts intercalary in origin, spherical to ovoid, 3–14 μm in diameter; spores somewhat larger than the vegetative cells, spherical, ovoid, or cylindrical with rounded ends, 6–20 μm in diameter, 8–40 μm long, single or in a series on either side of a heterocyst, the walls often becoming yellowish or brownish.

Of worldwide distribution in both freshwater and the sea. This species is most common in the intertidal zone where it is usually blue-green in color and in the form of free trichomes or a thin layer of trichomes in a soft, gelatinous polysaccharide. Spores are usually absent. These plants have been referred to species of the genus *Hormothamnion* in the past.

Bory 1822–1831, p. 29; Fries 1835, p. 330 (as "Anabaena"); Bornet and Flahault 1888, p. 233 (as "Anabaena"); Tilden 1910, p. 205, pl. 10, fig. 13 (as *Hormothamnion enteromorphoides* Grunow); Drouet 1978, p. 156, fig. 32–42.

Nostoc

Trichomes much constricted at the cross walls, moniliform, straight, curved, or loosely spiraled; vegetative cells discoid, spherical, ovoid, or cylindrical, isodiametric or shorter or longer than the diameter; the terminal cells discoid, spherical, ovoid, or cylindrical; sheaths usually thin, colorless, or fused to a gelatinous matrix, but apparently naked trichomes are common in marine habitats.

Nostoc spumigena (Mertens) Drouet. Trichomes 3–20 μm in diameter, much constricted at the nodes, moniliform, usually blue-green, yellow-green or olive, but sometimes brown, red, or violet, straight, curved, or spiraled, of indeterminate length fragmenting as a result of death of an intercalary cell, the maturation of a spore, or secretion of polysaccharide between two cells. Vegetative cells disk-shaped, compressed-spherical, or ovate-cylindrical; the terminal cell hemispherical or sometimes almost spherical; heterocysts intercalary, disk-shaped or compressed-spherical,

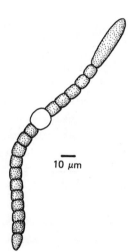

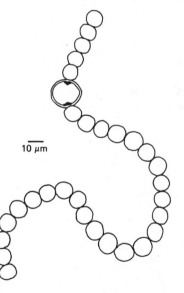

Figure 31. *Anabaina oscillarioides.* A trichome showing a spore, a heterocyst, and a slightly conical tip cell. The line is 10 microns.

Figure 32. *Nostoc spumigena.* A trichome showing moniliform cells and spherical tip cells. The line is 10 microns. From Humm (1979), The Marine Algae of Virginia, with permission of the University Press of Virginia.

3-20 μm in diameter; spores disk-shaped to spherical, intercalary, and formed progressively in a contiguous series. Sheaths may be lacking or distinct and hyaline, often becoming yellow or brownish, or they may be coalesced into a layer in which the trichomes are embedded.

Nostoc spumigena is most often encountered in shallow brackish and marine waters, and it is sometimes common in the plankton at which time the trichomes may have pseudovacuoles.

Mertens in Jürgens 1816–1822, p. 4; Bornet and Flahault 1888, p. 245 (as *Nodularia spumigena*); Drouet 1968, p. 122, fig. 24–25.

Family Stigonemataceae

Filaments branched, the lateral branches often more slender than the primary filaments. Heterocysts produced and, in some genera, spores. In the two marine genera, the trichomes are uniseriate but in some fresh water genera the trichomes are pluriseriate.

KEY TO THE GENERA AND SPECIES

Plants boring into shells or other forms of limestone
                                        Mastigocoleus testarum

Plants forming firm-gelatinous spheres or irregular masses in the intertidal zone along outer beaches   *Brachytrichia quoyi* (p. 92)

Mastigocoleus

Plants penetrating shells and other forms of calcium carbonate branched, the branches are uniseriate, of two kinds, one cylindrical, the other flagelliform and tapering to a hairlike tip. Heterocysts terminal on short cylindrical branches or sometimes intercalary; spores not produced. Extracellular polysaccharides forming a distinct sheath. One species.

Mastigocoleus testarum Lagerheim.   Plants producing a dense mass of anastomosing filaments within the surface layer of mollusc and barnacle shells, calcareous annelid tubes and bryozoan tests, and other forms of limestone. Trichomes 3.5–6.0 μm in diameter; the cells cylindrical or with small constrictions at the nodes, mostly isodiametric. Sheaths thin; and colorless. Heterocysts somewhat larger than the vegetative cells, spherical, 6–18 μm in diameter, at the end of short, cylindrical branches or intercalary.

Perkins and Tsentas (1967) found *M. testarum* to be the most abundant bluegreen penetrating the experimental calcareous material that they placed at intervals from the intertidal zone to a depth of 30 m, but especially from depths of 5–22 m.

Lagerheim 1886, p. 65, pl. 1, fig. 1–13; Bornet and Flahault 1887, p. 54; 1889, p. CLXII, pl. 10, fig. 4; Tilden 1910, p. 237, pl. 14, fig. 12; Taylor 1928, p. 49, pl. 2, fig. 2; Newton 1931, p. 41, fig. 27; Fremy 1934, p. 191, pl. 62, fig. 4; Lindstedt 1943, p. 32, pl. 3, fig. 1–2; Desikachary 1959, p. 575, pl. 122, fig. 1–5; Cocke 1967, p. 146, fig. 291; Hamm and Humm 1976, p. 212.

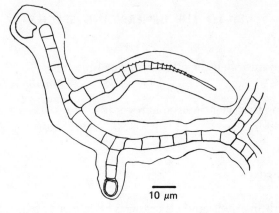

10 µm

Figure 33. *Mastigocoleus testarum.* A filament showing branching, including a flagelliform branch and a branch with a terminal heterocyst. The line is 10 microns. From Humm (1979), The Marine Algae of Virginia, with permission of the University Press of Virginia.

Brachytrichia

Trichomes embedded in a gelatinous matrix that is spherical or irregular in shape and usually attached to a solid object in the intertidal zone or below, sometimes epiphytic, solid at first, becoming hollow with age. Trichomes branched, often in a V-shape, some tapering to a thin hairlike tip, others increasing in diameter toward the apex, heterocysts intercalary; spores not produced. One species.

Brachytrichia quoyi (C. Agardh) Bornet and Flahault.   Plants forming a flattened crust or spherical to irregular, firm gelatinous masses a few millimeters to several centimeters in diameter, often folded and becoming hollow with age. Attached to rocks, wood, larger algae, or seagrass leaves in the intertidal zone along open-sea beaches where the salinity is consistently high and the water clear.

Trichomes 4–5 µm in diameter in the basal layer, 6–9 µm in diameter in the upper branches, some of which taper to 1 µm in diameter at the tip; upper branches parallel or radially arranged in the gelatinous matrix. Heterocysts 8–12 µm in diameter or more, spherical, oval, or subglobose.

Known in North America from New England to the West Indies and Gulf of Mexico, and along the entire Pacific coast, coastal waters of Europe, the Mediterranean Sea, and the Indian Ocean. It is especially common on pil-

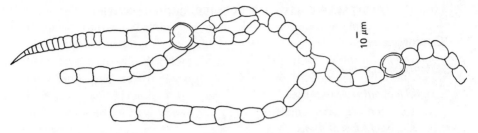

Figure 34. *Brachytrichia quoyi.* A filament showing branching, a tapering tip, and inter-calary heterocysts. The line is 10 microns. From Humm (1979), The Marine Algae of Virginia, with permission of the University Press of Virginia.

ings, breakwaters, and rocks along the coast of South Carolina to mid-Florida.

Agardh 1824, p. 22 (as *Nostoc quoyi*); Farlow 1881, p. 39 (as *Hormactis quoyi* [C. Agardh] Bornet); Lagerheim 1886, p. 65, pl. 1; Bornet and Flahault 1886, p. 373; Tilden 1910, p. 294, pl. 20, fig. 18; Geitler 1932, p. 554, fig. 347–8; Fremy 1936, p. 37, fig. 8 (as *B. balani* Bornet and Flahault); Linstedt 1943, p. 32, pl. 3, fig. 1–2; Cocke 1967, p. 172, fig. 326; Blackwelder 1975, p. 270.

SYSTEMATIC LIST OF SPECIES, sensu DROUET

Coccogonales

Chroococcaceae

Coccochloris stagnina
Coccochloris peniocystis
Coccochloris elabens
Johannesbaptistia pellucida
Agmenellum quadruplicatum
Agmenellum thermale
Gomphosphaeria aponina
Anacystis marina
Anacystis montana
Anacystis aeruginosa
Anacystis dimidiata

Chamaesiphonaceae

Entophysalis deusta
Entophysalis endophytica
Entophysalis conferta

Hormogonales

Oscillatoriaceae
Spirulina subsalsa
Oscillatoria erythraea
Oscillatoria lutea
Oscillatoria submembranacea
Schizothrix calcicola
Schizothrix mexicana
Schizothrix tenerrima
Schizothrix arenaria
Arthrospira neapolitana
Porphyrosiphon miniatus
Porphyrosiphon kurzii

Porphyrosiphon notarisii
Microcoleus vaginatus
Microcoleus lyngbyaceus

Nostocaceae

Calothrix crustacea
Scytonema hofmannii
Anabaina oscillarioides
Nostoc spumigena

Stigonemataceae

Mastigocoleus testarum
Brachytrichia quoyi

APPENDIX

The older names for the bluegreen algae (synonyms, *sensu* Drouet) are often useful in that they may represent growth forms or ecophenes of Drouet's names. For this reason, and also to provide the student with a convenient means of comparing the older bluegreen taxonomy with the concepts of Drouet, keys and descriptions of the majority of the bluegreen algae reported from marine habitats are provided in this appendix. In Table 1, which follows, is an alphabetical list of Drouet's valid names for those species known to grow in marine habitats. Under each is a list of older names or synonyms that have been reported from marine habitats. They are treated in the keys and descriptions in this appendix.

At the end of the appendix is an alphabetical list of these older names followed by their valid name as applied by Drouet (Table 2).

The number of older names for Drouet's species varies considerably from taxon to taxon, as it is related to the degree of variability with environmental conditions of each as well as its abundance and distribution. In the case of those reported from the sea, the number of synonyms ranges from one (*Anacystis dimidiata*) to twenty-six (*Calothrix crustacea*).

Table 1. List of Species sensu Drouet, and the Synonyms of Each that are Treated in the Appendix

Drouet's Valid Name	Older Name or Synonyms
Agmenellum quadruplicatum	*Merismopedia punctata forma minor* *Merismopedia warmingiana*
Agmenellum thermale	*Merismopedia elegans* variety *marina* *Merismopedia glauca forma mediterranea* *Merismopedia sabulicola*
Anabaina oscillarioides	*Anabaena inaequalis* *Anabaena torulosa* *Anabaena variabilis* *Hormothamnium enteromorphoides* *Hormothamnium solutum* *Nodularia hawaiiensis*
Ancystis dimidiata	*Chroococcus turgidus*
Anacystis marina	*Aphanocapsa marina*

Table 1. *Continued*

Drouet's Valid Name	Older Name or Synonyms
Anacystis montana forma montana	*Entophysalis magnoliae*
Arthrospira neapolitana	*Oscillatoria brevis*
Calothrix crustacea	*Calothrix aeruginea*
	Calothrix confervicola
	Calothrix consociata
	Calothrix contarenii
	Calothrix crustacea
	Calothrix fasciculata
	Calothrix fusco-violacea
	Calothrix parasitica
	Calothrix prolifera
	Calothrix pulvinata
	Calothrix scopulorum
	Calothrix vivipara
	Dichothrix bornetiana
	Dichothrix fucicola
	Dichothrix penicillata
	Dichothrix rupicola
	Fremyella aeruginea
	Fremyella grisea
	Fremyella longifila
	Richelia intracellularis
	Rivularia atra
	Rivularia bornetiana
	Rivularia coadunata
	Rivularia nitida
	Rivularia polyotis
Coccochloris stagnina	*Aphanothece pallida*
	Chroococcus minutus
Entophysalis conferta	*Dermocarpa biscayensis*
	Dermocarpa leibleiniae
	Dermocarpa minima
	Dermocarpa olivacea
	Dermocarpa prasina
	Dermocarpa rosea
	Dermocarpa smaragdina
	Dermocarpa solitaria

Table 1. *Continued*

Drouet's Valid Name	Older Name or Synonyms
Entophysalis conferta (*continued*)	*Dermocarpa violacea*
	Oncobyrsa marina
	Pleurocapsa amethystea
	Xenococcus acervatus
	Xenococcus schousboei
Entophysalis deusta	*Entophysalis granulosa*
	Entophysalis violacea
	Gloeocapsa crepidinum
	Hyella caespitosa
	Pleurocapsa crepidinum
	Pleurocapsa fuliginosa
Microcoleus lyngbyaceus	*Hydrocoleum cantharidosmum*
	Hydrocoleum comoides
	Hydrocoleum glutinosum
	Hydrocoleum lyngbyaceum
	Lyngbya aestuarii
	Lyngbya confervoides
	Lyngbya majuscula
	Lyngbya semiplena
	Oscillatoria bonnemaisonii
	Oscillatoria corallinae
	Oscillatoria margaritifera
	Oscillatoria miniata
Microcoleus vaginatus	*Oscillatoria amoena*
	Phormidium autumnale
Nostoc spumigena	*Nodularia harveyana*
	Nodularia spumigena
	Nostoc entophytum
	Nostoc linckia
Oscillatoria erythraea	*Trichodesmium contortum*
	Trichodesmium erythraeum
	Trichodesmium thiebautii
Oscillatoria submembranacea	*Phormidium penicillatum*
	Phormidium submembranaceum
	Symploca atlantica
Porphyrosiphon notarisii	*Oscillatoria nigro-viridis*
	Oscillatoria subuliformis

Table 1. *Continued*

Drouet's Valid Name	Older Name or Synonyms
Schizothrix arenaria	*Microcoleus chthonoplastes*
	Oscillatoria laetevirens
	Oscillatoria salinarum
	Phormidium subuliforme
	Schizothrix longiarticulata
	Symploca laete-viridis
Schizothrix calcicola	*Amphithrix janthina*
	Amphithrix violacea
	Lyngbya epiphytica
	Lyngbya rivulariarum
	Phormidium crosbyanum
	Phormidium fragile
	Phormidium persicinum
	Plectonema battersii
	Plectonema calothrichoides
	Plectonema nostocorum
	Plectonema terebrans
	Schizothrix lacustris
Schizothrix mexicana	*Lyngbya gracilis*
	Lyngbya meneghiniana
	Lyngbya sordida
	Phormidium spongeliae
	Symploca hydnoides
Schizothrix tenerrima	*Microcoleus tenerrimus*
	Oscillatoria acuminata
Spirulina subsalsa	*Spirulina major*
	Spirulina meneghiniana
	Spirulina nordstedtii
	Spirulina subsalsa
	Spirulina subtilissima
	Spirulina tenerrima
	Spirulina versicolor
Scytonema hofmannii	*Calothrix pilosa*
	Scytonema fuliginosum

KEY TO THE OLDER GENERA OF THE CYANOPHYTA

1 Cells spherical, ovate, or cylindrical, secreting extracellular polysaccharide on all wall surfaces; usually in colonies (Coccogonales)   2

1 Cells forming filaments, not secreting extracellular polysaccharide through the cross walls but only through the side walls (Hormogonales)   12

2 Cells or colonies directly and firmly attached and sometimes penetrating a substratum   3

2 Cells or colonies free or only loosely attached by adhesion of the sheath   8

3 Cells in radial rows; colonies cushionlike, hard, verrucose
Oncobyrsa (p. 111)

3 Cells not arranged in radial rows   4

4 Cells penetrating limestone or wood in which they form pseudofilaments that are often branched   *Hyella* (p. 118)

4 Cells not, or only slightly, penetrating limestone or wood   5

5 Cells in short, pseudofilamentous rows surrounded by an elliptical sheath   *Entophysalis* (p. 116)

5 Cells irregular in arrangement, not in rows   6

6 Plants reproducing by "endospores" only   *Dermocarpa* (p. 111)

6 Reproduction by cell division and "endospores"   7

7 Colonies hemispherical to spherical, of several layers of cells
Pleurocapsa (p. 114)

7 Colonies flat, usually of one layer of cells   *Xenococcus* (p. 118)

8 Cells in one layer embedded in a flat polysaccharide disk in which they are check-rowed   *Merismopedia* (p. 106)

8 Cells single or irregularly arranged in the sheath   9

9 Cells spherical, each with a definite sheath or in twos, fours, or eights   *Chroococus* (p. 105)

 9 Cells in larger groups in a common gelatinous sheath   10

10 Cells spherical (or adjacent faces flattened)   11

10 Cells oblong, dividing in a plane perpendicular to the long axis
                                                    Aphanothece (p. 109)

11 Individual cell sheaths remaining distinct in the colony
                                                    Gloeocapsa (p. 110)

11 Individual cell sheaths not remaining distinct in the common sheath
                                                    Aphanocapsa (p. 110)

12 Trichomes spiral, no cross walls   *Spirulina* (p. 119)

12 Trichomes with cross walls, rarely spiral   13

13 Plants strictly marine and planktonic, trichomes in bundles forming
    visible colonies   *Trichodesmium* (p. 123)

13 Plants benthic or epiphytic, only occasionally planktonic   14

14 Trichomes without heterocysts   15

14 Trichomes with heterocysts   23

15 Trichomes without a visible sheath or gelatinous matrix
                                                    Oscillatoria (p. 124)

15 Trichomes with a sheath or in a common gelatinous matrix   16

16 Trichomes single within the sheath   17

16 Trichomes several to many within a sheath   21

17 Trichomes without false branching   18

17 Trichomes with false branching   20

18 Plants with a horizontal layer and erect filaments that taper to a fine
    point   *Amphithrix* (p. 146)

18 Filaments not tapering to a fine or hairlike point   19

19 Sheaths distinct, not confluent or fused   *Lyngbya* (129)

19 Sheaths confluent or fused forming a matrix in which the trichomes
    are embedded   *Phormidium* (p. 134)

20 Filaments arising from a prostrate base in wicklike bundles; false branches solitary   *Symploca* (p. 138)

20 Basal layer absent; false branches single or in pairs
                                                              Plectonema (p. 140)

21 Trichomes few within the sheath   22

21 Trichomes many within the sheath   *Microcoleus* (144)

22 End wall of tip cell thickened; granules on the cross walls
                                                              Hydrocoleum (p. 142)

22 End wall of tip cell thin; no granules against the cross walls
                                                              Schizothrix (p. 145)

23 Trichomes tapering to a hairlike tip   24

23 Trichomes without a hairlike tip   26

24 Without false branches, or false branches free from each other
                                                              Calothrix (p. 147)

24 False branches not all separate from each other   25

25 Groups of false branches remaining within the original sheath
                                                              Dichothrix (155)

25 Entire plant inside a common gelatinous matrix   *Rivularia* (p. 157)

26 False branches present   *Scytonema* (p. 160)

26 Filaments without false branches   27

27 Filaments attached at base, erect or decumbent   *Fremyella* (p. 153)

27 Filaments not attached at base, prostrate   28

28 Trichomes with a sheath; plants often forming a thin layer or dendroid mass   *Hormothamnium* (p. 160)

28 Trichomes without a sheath or with a matrix of fused sheaths   29

29 Plants endophytic in diatoms   *Richelia* (p. 153)

29 Plants not endophytic   30

30 Trichomes within a spherical or irregular gelatinous matrix
                                                              Nostoc (p. 164)

30   Trichomes free or aggregated   31

31   Cells disk-shaped, shorter than their diameter   *Nodularia* (p. 166)

31   Cells mostly isodiametric; trichomes often aggregated in a layer
                                                                                    Anabaina (p. 161)

Chroococcus

Plants are in the plankton or attached and forming a crust or gelatinous masses; the cells are spherical or with adjacent faces flattened following division, each surrounded by a sheath, solitary, in twos, fours, or eights but not in many-celled colonies.

Two species have been reported many times from marine habitats of Europe and North America.

KEY TO THE SPECIES OF CHROOCOCCUS

Cells 6–9 μm in diameter   *C. minutus*

Cells 15–40 μm in diameter   *C. turgidus*

Chroococcus minutus (Kützing) Nägeli.   Cells occuring singly or in twos (sometimes in fours), spherical to ovoid, with adjacent faces flattened if recently divided; 6–9 μm in diameter, 10–13 μm long. Individual cell sheaths not always distinguishable, the sheaths not showing concentric layers.

Kützing 1843, p. 168; 1849, p. 197, pl. 5 (as *Protococcus minutus*); Nägeli 1849, p. 46; Tilden 1910, p. 7, pl. 1, fig. 5; Geitler 1932, p. 232; Lindstedt 1943, p. 19, pl. 1, fig. 11.

Chroococcus turgidus (Kützing) Nägeli.   Cells solitary or in twos or fours within a stratified, hyaline sheath, 13–25 μm in diameter, spherical, but with adjacent faces flattened for some time after division.

This form of *A. dimidiata* is the largest species of the old genus *Chroococcus*. It is common in shallow brackish and marine waters on bottom sediments or clinging to larger algae, invertebrates, or other solid surfaces. It is easily overlooked because of the fact that it does not accumulate in visible masses.

Kützing 1849, p. 5, pl. 6 (as *Protococcus turgidus*); Nägeli 1849, p. 46; Tilden 1910, p. 5, pl. 1, fig. 3; Newton 1931, p. 3, fig. 1; Geitler 1932, p. 228, fig. 110; Lindstedt 1943, p. 20, pl. 1, fig. 12–13; Chapman 1956, p. 349, fig. 2a.

Merismopedia

Colonies tending to be rectangular and flat, the cells in a single layer and in check rows in the gelatinous matrix, spherical or with adjacent faces flattened after division. Cell division is synchronous and alternately in two planes at right angles to each other. The cell arrangement in older colonies may become irregular.

106

The genus is well known in freshwater, but it is also common in marine habitats. Marine phycologists, however, have usually overlooked it. Lindstedt (1943) is an exception. He records four species from marine habitats of the west coast of Sweden and *Holopedia sabulicola* Lagerheim (*Merismopedia sabulicola* Lagerheim), a fifth, transferred to *Holopedia* because the cells were not in regular rows.

KEY TO THE SPECIES OF MERISMOPEDIA

1 Cells in irregular arrangement in one plane   *M. sabulicola*

1 Cells in check rows in the colonies   2

2 Colonies with 64 or fewer cells   3

2 Colonies with more than 64 cells

M. glauca forma mediterranea

3 Cells 0.8–1.5 μm in diameter   *M. warmingiana* (p. 109)

3 Cells over 1.5 μm in diameter   4

4 Cells 1.5–2.5 μm in diameter   *M. punctata forma minor* (p. 109)

4 Cells 6–8 μm in diameter   *M. elegans* variety *marina* (p. 109)

Merismopedia sabulicola Lagerheim.   Cells in one plane in the colony, but not in rows, spherical to ovate, 3–4 μm in diameter, 5–6 μm long; the contents without granules, bluegreen.

This form of *Agmenellum thermale* was found by Lindstedt in intertidal sand at several localities along the west coast of Sweden.

Lagerheim 1883, p. 43; Geitler 1930–1932, p. 267, fig. 131; Lindstedt 1943, p. 14, pl. 1, fig. 3–4 (as *Holopedia sabulicola*).

Merismopedia glauca (Ehrenberg) Nägeli *forma mediterranea* Nägeli.   Colonies squarish or rectangular, consisting of 64 or more cells that are spherical to ovate, 3–6 μm in diameter and in more or less regular check rows.

Forma mediterranea is known from the Mediterranean Sea and the west coast of Sweden. *M. glauca* has been reported from the Isle of Cumbrae, Scotland.

Drouet and Daily (1956) regard *M. glauca* as a synonym of *A. quadruplicatum*, but *forma mediterranea* as a synonym of *A. thermale*.

Nägeli 1849, p. 56, fig. 1D; Newton 1931, p. 8, fig. 7; Fremy 1934, p. 7; fig. 7; Fremy 1934, p. 7; Lindstedt 1943, p. 11 pl. 1, fig. 1.

Merismopedia warmingiana Lagerheim ex Forti.   Colonies squarish
or rectangular, small, consisting of 4–32 cells (usually 4–16) that are
0.8–1.5 μm in diameter, spherical, in check rows, bluegreen in color.

Recorded from several coastal habitats in intertidal sand along the west
coast of Sweden. This species is a synonym of *A. quadruplicatum* (Drouet
and Daily 1956).

Lagerheim in Forti 1907, p. 109; Lindstedt 1943, p. 12, pl. 1, fig. 2.

Merismopedia punctata Meyen *forma minor* Lagerheim. Colonies
squarish or rectangular, small, consisting of 4–64 cells (usually 4–16) that
are 1.5–2.5 μm in diameter, spherical to ellipsoidal, the protoplasm
bluegreen, often with fine granules.

This form of *A. quadruplicatum, sensu* Drouet and Daily (1956) is
known from the Baltic Sea near Kiel and from the west coast of Sweden
and of Denmark.

Meyen 1839, p. 67; Geitler 1930–1932, p. 263, fig. 129c; Lindstedt
1943, p. 13, pl. 1, fig. 6.

Merismopedia elegans A. Braun variety *marina* Lagerheim.   Colonies
squarish or rectangular small, of 8–64 cells in check rows or somewhat irregular in arrangement in the gelatinous matrix, 6–8 μm in diameter,
6–10 μm long; spherical to ovate, bluegreen, the protoplasm granular.

This form of *A. thermale, sensu* Drouet and Daily (1956) is widely
known in freshwater. Variety *marina* is apparently known only from the
west coast of Sweden.

Braun in Kützing 1849, p. 472; Geitler 1930–1932, p. 265; Lindstedt
1943, p. 13, pl. 1, fig. 5; Cocke 1967, p. 18, fig. 46.

Aphanothece

Colonies irregular in shape, but tending to be spherical, the cells ovate to
oblong, dividing in a plane at right angles to the long axis, irregularly arranged in the gelatinous matrix.

Aphanothece pallida (Kützing) Rabenhorst.   Colonies soft-gelatinous
and hyaline, the cells ovate to oblong-elliptical or cylindrical, 3–8 μm in
diameter, 5–24 μm long, to three times as long as the diameter.

This form of *Coccochloris stagnina sensu* Drouet and Daily (1956), was found on rocks among brown algae on the coast of Dorset, England (Newton 1931).

Rabenhorst 1865, p. 64; Newton 1931, p. 7, fig. 6A-B; Geitler 1932, p. 171, fig. 78.

Gloeocapsa

Plants in the form of small colonies usually containing 2–8 cells, the large colonies often with secondary colonies, cells spherical, mostly 2–10 μm in diameter. Nannocytes and spores have been reported (Desikachary 1959, p. 111).

One species has been found repeatedly in marine habitats along the coasts of Europe and North America.

Gloeocapsa crepidinum (Rabenhorst) Thuret.   Cells enclosed in a thick, gelatinous, hyaline sheath that is usually ellipsoid, oval, or spherical in which the cells are solitary, in twos or fours, or larger masses consisting of secondary colonies; cells spherical, 4–8 μm in diameter; cell division in three directions.

This species is an unattached or loosely adherent form of *Entophysalis deusta,* according to Drouet and Daily (1956) so that records of nannocytes is not surprising.

Thuret in Bornet and Thuret 1876–1880, p. 1, pl. 1, fig. 1–3; Tilden 1910, p. 20, pl. 1, fig. 28; Newton 1931, p. 4, fig. 3; Geitler 1932, p. 190, fig. 85–86; Lindstedt 1943, p. 21, pl. 2, fig. 1; Desikachary 1959, p. 117, pl. 27, fig. 4–5.

Aphanocapsa

Cells enclosed within a formless gelatinous matrix that is hyaline and soft; the individual sheaths not visible or indistinct; cells spherical, irregularly distributed, or in pairs following division.

Aphanocapsa marina Hansgirg.   Colonies small, soft-gelatinous, often clinging to larger algae or solid surfaces, or planktonic; cells spherical to ovate, 1–2 μm in diameter.

This species was described from a rock pool in the splash zone on the coast of Norway. It is a synonym of *Anacystis marina* sensu Drouet and Daily (1956). It has also been called *Anacystis nidulans* (Drouet and Daily 1952).

Newton (1931) reports *Aphanocapsa marina* from larger marine algae in the British Isles but gives the cell dimensions as 5 μm in diameter, 6.0–7.5 μm long. This material was perhaps another species of *Aphanocapsa*.

Hansgirg in Foslie 1890–1891, p. 169; Newton 1931, p. 3, fig. 2; Drouet and Daily 1956, fig. 376 (photograph of the isotype).

Oncobyrsa

Forming a layer or stratum that is hard and leathery on larger algae, the cells spherical to elongate and usually in radial rows.

Oncobyrsa marina (Grunow) Rabenhorst. Colonies firm, rugose, growing upon larger algae; sheaths thick and firm, cells spherical to ovate, often in radial rows, 1.5–2.0 μm in diameter.

This species is a small-celled form of *Entophysalis conferta sensu* Drouet and Daily (1956). It has been reported from marine habitats in Florida, Bermuda, the Mediterranean, the British Isles and Africa.

Rabenhorst 1865, p. 68; Howe 1920, p. 620; Taylor 1928, p. 42; Newton 1931, p. 5, fig. 4.

Dermocarpa

Plants epiphytic, forming small colonies or expanded groups or layers usually one cell in thickness, the cells spherical, ovate, pyriform, or cylindrical; cells often dividing repeatedly into numerous small "endospores" or "gonidia" with ultimate dissolution of the retaining sheath at the top.

All species of *Dermocarpa* listed below are regarded as forms of *Entophysalis conferta* by Drouet and Daily (1956), except *D. violacea* which is a synonym of *E. deusta*.

KEY TO THE SPECIES OF DERMOCARPA

1 Cells oval to pyriform, narrowed at base but not forming a distinct stalk   **2**

1 Cells or sheath contracted at base into a stalk   **6**

2 Cell contents violet or rose   **3**

2 Cell contents green or a shade of green   **4**

3 Cells violet, 8–28 μm in diameter   *D. violacea*

3 Cells rose, 4–5 μm in diameter   *D. rosea* (p. 113)

4 Cells solitary or in a small group, 9–14 μm in diameter
   D. solitaria (p. 113)

4 Cells numerous in colony, closely packed, often angular by mutual pressure   **5**

5 Cells 4–6 μm in diameter, claviform   *D. biscayensis* (p. 113)

5 Cells 4–24 μm in diameter, cylindrical to oblong
   D. prasina (p. 113)

6 Cells usually solitary   *D. minima* (p. 114)

6 Cells forming colonies   **7**

7 Cells nearly spherical, 18–24 μm in diameter
   D. leibleiniae (p. 114)

7 Cells elongate, often pyriform   **8**

8 Cells 8–11 μm in diameter   *D. smaragdina* (p. 114)

8 Cells 10–17 μm in diameter   *D. olivacea* (p. 114)

Dermocarpa violacea Crouan. Colonies forming small reddish-violet patches on the host, 3–5 mm in diameter; cells ovate to pyriform, 8–28 μm in diameter, the sheaths thin.

Known from the British Isles and the Atlantic and Pacific coasts of North America.

Crouan 1858, p. 70, pl. 3, fig. 2; Collins 1900, p. 41; Tilden 1910, p. 53, pl. 3, fig. 19–21; Newton 1931, p. 10; Geitler 1932, p. 397, fig. 221c–e.

Dermocarpa rosea (Reinsch) Batters. Colonies forming patches on the host that are typically 2–5 cm in diameter; the sheath thick, hyaline; the individual sheaths distinct; cells 4–5 μm in diameter, ovoid to elliptical, the cell contents rose red.

Known from the Atlantic coast of Canada and the British Isles.

Batters 1889, p. 141; Tilden 1910, p. 53, pl. 3, fig. 16–18; Newton 1931, p. 10; Geitler 1932, p. 396, fig. 221a–b.

Dermocarpa solitaria Collins and Hervey. Cells occuring singly or in few-celled colonies scattered on the host plants, the base flattened but the cells becoming broader at the top; sheaths thick, hyaline, cells 1–14 μm in diameter, 20–60 μm tall; dividing unequally, the upper daughter cell often subdividing into 8–12 endospores that are 5–6 μm in diameter.

Described from Bermuda but also recorded for Dry Tortugas, Florida.

Collins and Hervey 1917, p. 17; Taylor 1928, p. 41, pl. 1, fig. 1.

Democarpa biscayensis Sauvageau. Colonies small, up to about 1 mm in diameter, rounded, the margins often lobed or irregular, the cells claviform, closely packed and becoming angular by mutual pressure, 4–6 μm in diameter. Endospores not reported.

Described from the Bay of Biscay and recorded from the Canary Islands and the coast of Spain.

Sauvageau 1895, p. 7, pl. 7, fig. 2–3; Geitler 1932, p. 397; Fremy 1934, p. 60, pl. 17, fig. 2.

Dermocarpa prasina (Reinsch) Bornet. Colonies cushionlike, small, with thin sheaths, cells wedge-shaped or oblong from mutual pressure, 4–24 μm in diameter, 15–30 μm tall; "endospores" seriate or multiseriate within the parent cell sheath.

Known from Europe and the Atlantic and Pacific coasts of North America.

Bornet in Bornet and Thuret 1876–1880 (1880), p. 73, pl. 26, fig. 6–9; Tilden 1910, p. 52, pl. 3, fig. 13–15; Taylor 1928, p. 41, pl. 1, fig. 8–9; Newton 1931, p. 10, fig. 7A–C.

Dermocarpa minima (Geitler) Drouet.   Cells often solitary but also in groups of a few, with a slender gelatinous stalk at the base, 5–6 μm in diameter; "endospores" 8–16 in number.

Known from California and the Canary Islands.

Geitler 1932, p. 392, fig. 214b–c; Fremy 1934, p. 58, pl. 15, fig. 4; 1936, p. 13.

Dermocarpa leibleiniae (Reinsch) Bornet.   Cells pyriform, ovoid, or elliptical, 18–24 μm in diameter, solitary or in small, close groups, forming a thin stalk at the base; sheaths thick, often laminated.

Known from Bermuda, the Canary Islands, and Britain.

Bornet in Bornet and Thuret 1878–1880, p. 73; Tilden 1910, p. 55, pl. 3, fig. 28; Newton 1931, p. 9; Fremy 1936, p. 16.

Dermocarpa smaragdina (Reinsch) Tilden.   Cells broadly pyriform, 8.5–11 μm in diameter, 16–33 μm tall, the base narrowed to a hyaline stalk about 2 μm wide.

Known as an epiphyte of larger algae along the Atlantic coast of North America from New England to Labrador.

Reinsch 1875, p. 16, pl. 25, fig. 4 (as *Sphaenosiphon smaragdinus*); Farlow 1881, p. 61; Tilden 1910, p. 54, pl. 3, fig. 24–25; Geitler 1932, p. 401, fig. 227.

Dermocarpa olivacea (Reinsch) Tilden.   Colonies small and hemispherical; sheaths thick, lamellose, cells broadly pyriform, ovate, or spherical, the adjacent faces of newly divided cells flattened, 9.5–17 μm in diameter, 12–25 μm long, endospores numerous and spherical.

Known from the eastern maritime provinces of Canada, British Isles, Canary Islands, Adriatic Sea, Ceylon, India.

Reinsch 1875, p. 17, pl. 27, fig. 2 (as *Sphaenosiphon olivaceus*); Tilden 1910, p. 55, pl. 3, fig. 26–27; Fremy 1933, p. 61, pl. 16, fig. 7; Desikachary 1959, p. 174, pl. 33, fig. 13–14.

Pleurocapsa

Colonies hemispherical to subspherical, of several layers of cells, forming an encrusting growth with extensions into the substrate on the lower side, the upper surface often with uniseriate or multiseriate branches.

KEY TO THE SPECIES OF PLEUROCAPSA

1   Cells 5–20 μm in diameter, the contents yellowish to dull violet
                                               P. fuliginosa

1   Cells 10–15 μm in diameter   **2**

2   Cell contents violet; plants epiphytic   *P. amethystea*

2   Cell contents dull blue or slate; plants on stones, wood
                                             P. crepidinum

Pleurocapsa fuliginosa Hauck. Colonies blackish, 50–100 μm in diameter; the cells 5–20 μm in diameter, solitary, or in twos or fours, embedded in a firm sheath.

On stones, shells, and pilings in the intertidal zone in Europe, the British Isles, both coasts of North America.

Hauck 1885, p. 515; Tilden 1910, p. 48, pl. 3, fig. 2–3; Newton 1931, p. 10, fig. 9; Drouet and Daily 1956, fig. 189, photograph of the type specimen. (*Entophysalis deusta*).

Pleurocapsa amethystea Rosenvinge. Colonies epiphytic, minute, hemispherical to spherical, about 50 μm in diameter, and dark violet, cells 10–13 μm in diameter, spherical, or with adjacent faces flattened, solitary or in small groups, "endospores" 1–2 μm in diameter.

This name is a synonym of *Entophysalis conferta sensu* Drouet and Daily (1956). It is reported from the British Isles northward to the Arctic Sea and Greenland.

Rosenvinge 1893, p. 967, fig. 57; Tilden 1910, p. 48, pl. 3, fig. 4; Newton 1931, p. 10.

Pleurocapsa crepidinum Collins. Colonies on stones and wood, hemispherical to spherical, minute, slate blue, often in dense masses, cells to 15 μm in diameter, spherical, or with adjacent faces flattened, small "endospores" produced.

This form of *Entophysalis deusta sensu* Drouet and Daily was described from intertidal woodwork in New England.

Collins 1901, p. 136; Tilden 1910, p. 49; Geitler 1932, p. 355, fig. 188a–b.

Entophysalis

Cells spherical, each within a firm sheath, these associate into colonies of various shapes that are firm gelatinous or cartilaginous, growing upon stones, shells, submerged wood.

KEY TO THE SPECIES OF ENTOPHYSALIS

1   Cells 4–6 µm in diameter   *E. magnoliae*

1   Cells 2–5 µm in diameter   **2**

2   Colonies to 1 mm thick, surface rough, often dendroid

                                          E. granulosa

2   Colonies less than ½ mm thick, surface smooth   *E. violacea*

Entophysalis magnoliae Farlow.   Colonies consisting mostly of groups of 2–4 cells in individual sheaths that are massed together, cells 4–6 µm in diameter, dark purple; forming a thin gelatinous film on intertidal rocks in New England.

Drouet and Daily (1956) regard this species as a form of *Anacystis montana, forma montana.*

Farlow 1881, p. 29; Collins 1900, p. 41; Tilden 1910, p. 24.

Entophysalis granulosa Kützing.   Plants forming a brown or blackish expanded crust on rocks and wood in the intertidal or splash zone that is rough or even dendroid on the surface consisting of small colonies of cells in twos, fours, or eights each in a firm sheath, these fused to form the crust, cells 2–5 µm in diameter, spherical or with flattened adjacent faces.

This widely distributed plant is a form of *Entophysalis deusta* according to Drouet and Daily (1956).

Kützing 1843, p. 177, pl. 18, fig. 5; Tilden 1910, p. 24, pl. 1, fig. 33; Newton 1931, p. 11, fig. 10; Fremy 1934, p. 32, fig. 6.

Entophysalis violacea Collins.   Plants forming a thin, smooth crust on rocks in the intertidal zone consisting of colonies of two or four cells in individual sheaths that are fused into the crust and often arranged in longitudinal series, cells 2–3 µm in diameter, dark violet, and spherical, but elongate before division.

Described from rocks low in the intertidal zone in the Bahamas. This plant is another form of *Entophysalis deusta* (Drouet and Daily 1956).

Howe 1920, p. 619; Newton 1931, p. 10.

Hyella

Colonies of various sizes forming a radially expanding cushion on limestone or wood, the basal part of the colony forming pseudofilaments that are branched and penetrate the substrate and in which the cells are separated from each other by sheath material, the upper part of the colony of coccoid cells of various sizes; these often subdividing into numerous "endospores" held together for a time by the sheath of the original cell.

One species of this genus is cosmopolitan in marine habitats. It is included in *Entophysalis deusta* by Drouet and Daily (1956).

Hyella caespitosa Bornet and Flahault. With characteristics of the genus. These plants are on old shells and corals to which they lend a greenish-black or green tinge, the superficial cells coccoid, the basal pseudofilaments penetrating limestone or wood as deeply as light permits. They are abundant along all seacoasts of the world, and are best seen after decalcification of bits of greenish limestone in dilute hydrochloric acid. The cells are not seriously altered by this treatment.

Bornet and Flahault 1888, p. 162; Tilden 1910, p. 51, pl. 3, fig. 9–11; Taylor 1928, p. 42, pl. 1, fig. 10–11; Newton 1931, p. 13, fig. 11; Lindstedt 1943, p. 27, pl. 2, fig. 7–8.

Xenococcus

Colonies flat to hemispherical or a curved disk, epiphytic, of one cell layer at first, later becoming several cells thick and forming short, erect protrusions, cells spherical, ovate, pyriform, often laterally flattened by mutual pressure, surface or marginal cells often dividing into numerous endospores.

The two species widely reported from marine habitats are forms of *Entophysalis conferta* (Drouet and Daily 1956), as are two other species not treated here, *X. chaetomorphae* Setchell and Gardner and *X. cladophorae* (Tilden) Setchell and Gardner, both reported from the Pacific coast of North America and from India (Desikachary 1959, pp. 182–183).

KEY TO THE SPECIES OF XENOCOCCUS

Colonies of a single layer of cells, endospores produced

$\qquad$ *X. schousboei*

Colonies becoming multilayered, endospores unknown

$\qquad$ *X. acervatus*

Xenococcus schousboei Thuret.   Plants in the form of solitary or scattered cells or closely grouped with confluent cells forming a layer on filamentous and other algae, cells mostly 4–9 μm in diameter, spherical, ovate, or pyriform, often with flattened adjacent faces where crowded; the cell layer becoming double or multiple with age, numerous "endospores" (to 32) may be produced by the upper cells. Cosmopolitan.

Thuret in Bornet and Thuret 1880, p. 74, pl. 26, fig. 1–2; Tilden 1910, p. 50, pl. 3, fig. 7; Drouet and Daily 1956, fig. 210, a photograph of the type.

Xenococcus acervatus Setchell and Gardner.   Plants in the form of a layer of cells on the surface of an algal host, later growing on top of the original layer and becoming double or multiple with age; cells spherical, pyriform, or angular by pressure, "endospores" not known.

Known from Europe, California, the Canary Islands, and Ceylon.

Setchell and Gardner in Gardner 1918b, p. 459, pl. 39, fig. 13; Fremy 1934, p. 43, pl. 8, fig. 6; Desikachary 1959, p. 182, pl. 31, fig. 28; Drouet and Daily 1956, fig. 212, a photograph of the type.

Spirulina

Tilden (1910) recognized nine species of the genus *Spirulina* that had been recorded for North America or Hawaii, separated by trivial differences that vary with the environment, not only from place to place but even within the microenvironment of a given plant mass. The basis for separation of *S. major* and *S. meneghiniana*, the degree of regularity of the spiral, is an extreme application of taxonomic trivia and at the same time a commentary on the older taxonomy of the bluegreens.

Of Tilden's nine species, seven (including *S. subsalsa, sensu* the older literature) have been recorded from marine environments and are treated in the key, descriptions, and habitat notes that follow.

Drouet (1968) recognized but one species, *Spirulina subsalsa.*

KEY TO SELECTED OLDER NAMES FOR
SPIRULINA SUBSALSA

1	Forming a loose spiral	2
1	Forming a tight spiral, at least in part	6
2	Trichomes 0.4 μm in diameter	*S. tenerrima*
2	Trichomes of greater diameter	3
3	Trichomes 0.6–0.9 μm in diameter	*S. subtilissima*
3	Trichomes of greater diameter	4
4	Trichomes 1–2 μm in diameter, 3–5 μm between turns	5
4	Trichomes 2 μm in diameter, 5 μm between turns, the spiral regular	*S. nordstedtii* (p. 122)
5	Spiral irregular	*S. meneghiniana* (p. 122)
5	Spiral regular	*S. major* (p. 122)
6	Spiral regular, the turns contiguous, plant mass purple	*S. versicolor* (p. 122)
6	Spiral somewhat irregular, the turns contiguous or nearly so, plant mass bluegreen or yellow-green	*S. subsalsa* (p 123)

Spirulina tenerrima Kützing. Trichomes 0.4 μm in diameter, in regular spirals that are 1.4–1.6 μm in amplitude and about 1 μm from each other, cell contents pale bluegreen.

On rocks in sheltered bays mixed with other bluegreen algae. Known from North America, Europe, and Africa.

Kützing 1843, p. 183; Gomont 1892, p. 252; Tilden 1910, p. 88; Fremy 1934, p. 132, pl. 31, fig. 21; Lindstedt 1943, p. 55, pl. 6, fig. 10–11; Cocke 1967, p. 34, fig. 72.

Spirulina subtilissima Kützing. Trichomes 0.6–0.9 μm in diameter with a regular spiral 1.5–2.5 μm in amplitude; the distance between the turns is 1.2–2.0 μm.

Epiphytic on marine algae; also known from inland waters with traces of salt and from sulfur springs. Of worldwide distribution.

Kützing 1843, p. 183; Gomont 1892, p. 252, pl. 7, fig. 30; Tilden 1910, p. 88, pl. 4, fig. 47; Fremy 1934, p. 132, fig. 31; Lindstedt 1943, p. 56, pl. 6, fig. 12–13; Drouet 1968, fig. 3, a drawing of part of the type specimen.

Spirulina nordstedtii Gomont.   Trichomes about 2 μm in diameter, forming a regular spiral of about 5 μm diameter; the distance between the turns is also about 5 μm.

Known from the coast of Maine and from freshwater in Europe.

Gomont 1892, p. 252; Collins 1900, p. 43; Tilden 1910, p. 88.

Spirulina meneghiniana Zanardini.   Trichomes 1.2–1.8 μm in diameter, forming a loose, irregular spiral of 3.2–5.0 μm amplitude; the distance between the turns 3–5 μm.

Known from brackish water, salt marshes, and rock tide pools in the splash zone. Cosmopolitan.

Gomont 1892, p. 250, pl. 7, fig. 28; Collins 1896, p. 1; 1900 p. 43; 1901, p. 289; Tilden 1910, p. 87, pl. 4, fig. 45; Geitler 1932, p. 928, fig. 593b.

Spirulina major Kützing.   Trichomes 1.2–1.7 μm in diameter in a somewhat loose but regular spiral having an amplitude of 2.4–4.0 μm and a distance between turns of 2.7–5.0 μm.

This form often occurs among larger algae as scattered trichomes or small plant masses. It is cosmopolitan in both salt- and freshwater.

Gomont 1892, p. 251, pl. 7, fig. 29; Tilden 1910, p. 87, pl. 4, fig. 46; Geitler 1932, p. 930, fig. 595; Newton 1931, p. 15, fig. 13; Fremy 1934, p. 131, fig. 31; Lindstedt 1943, p. 56, pl. 6, fig. 14.

Spirulina versicolor Cohn.   Trichomes 1.2–1.8 μm in diameter, forming a dense and regular spiral of 3.0–4.5 μm amplitude, the color violet-purple.

On mooring buoy, on larger algae, and shells in the North Atlantic, North Sea, and Adriatic Sea.

Gomont 1892, p. 253; Collins 1896, p. 458; 1900, p. 43; Tilden 1910, p. 89; Newton 1931, p. 15; Drouet 1968, fig. 7, a drawing of a trichome from the type specimen.

Spirulina subsalsa Oersted.   In the older literature this species name was applied to plants with trichomes 1–2 μm in diameter, having a somewhat irregular and dense spiral; that is, a section of the trichome with regular, often contiguous turns, and then a section with a loose spiral, these often alternating in the longer trichomes, both having an amplitude of 3–5 μm.

An abundant and widely distributed form in seawater and brackish water along both coasts of North America, including the Gulf of Mexico and the West Indies.

Oerstedt 1842, p. 17, pl. 7, fig. 4; Gomont 1892, p. 253, pl. 7, fig. 32; Tilden 1910, p. 89, pl. 4, fig. 49; Newton 1931, p. 15; Fremy 1934, p. 133, pl. 31; Lindstedt 1943, p. 57, pl. 7, fig. 3–5; Chapman 1961, p. 21, fig. 7; Cocke 1967, p. 33, fig. 67; Dawes 1974, p. 58, fig. 20.

Trichodesmium

Plants strictly planktonic and marine in the form of bundles of trichomes held together by soft, transparent sheath material, the colonies often scalelike or flattened; trichomes cylindrical, the tip cell a hemispherical or truncate-conical end. Plants of this genus often form great patches of discolored water in warm-temperate and tropical seas.

Ehrenberg described the genus and species *Trichodesmium erythraea* in 1830. The genus was accepted by Gomont in his monograph (1892). Geitler, in 1932, transferred the species of *Trichodesmium* to *Oscillatoria* as he felt that the difference between the two genera was trivial. In 1942, however, he went back to *Trichodesmium*.

In 1938 de Toni proposed the genus *Skujaella* to replace *Trichodesmium*, which he regarded as invalid. Drouet (1968) followed Geitler's earlier opinion and transferred all species of *Skujaella* to *Oscillatoria*, regarding them all as a single species because the differences, he felt, were products of the environment and there were series of intermediates between any two species. He placed them all in Ehrenberg's original species as *O. erythraea*.

KEY TO THE SPECIES OF TRICHODESMIUM

1   Trichomes straight, 7–11 μm in diameter   *T. erythraeum*

1   Trichomes curved or spirally bent   **2**

2   Trichomes 7–12 μm in diameter   *T. thiebautii*

2   Trichomes 16–25 μm in diameter   *T. contortum*

Trichodesmium erythraeum (Ehrenberg) Gomont. Trichomes straight, forming flakelike or scalelike bundles in the plankton, cells mostly isodiametric, or a little shorter than the diameter, 7–11 μm in diameter, 5.4–11 μm long, the apical cell truncate-conical or hemispherical, trichomes slightly constricted at the nodes, tapering slightly at the ends.
Gomont 1892, p. 196, pl. 5, fig. 27–30; Tilden 1910, p. 84, pl. 4, fig. 40; Geitler 1932, p. 968, fig. 617 a–d; Desikachary 1959, p. 245, pl. 42, fig. 1–2.

Trichodesmium thiebautii Gomont. Trichomes curved to spiral and forming colonies held together at the center with the ends free, 1–6 mm long, cells 7–16 μm in diameter, isodiametric or longer than wide, 6–26 μm long, the tip cell with thickened end wall, trichomes briefly attenuated at the tips, without constrictions at the cross walls.
Gomont 1892, p. 197, pl. 6, fig. 2–4; Tilden 1910, p. 84, pl. 4, fig. 41–42; Taylor 1928, p. 47, pl. 1, fig. 15; Geitler 1932, p. 907, fig. 617 c.

Trichodesmium contortum Wille. Trichomes loosely spiralled and forming twisted, ropelike colonies, the cells 16–25 μm in diameter and mostly isodiametric.
Brandt 1903, p. 18, fig. 14; Lemmermann 1905, p. 618; Tilden 1910, p. 85, pl. 4, fig. 43.

Oscillatoria

Trichomes without an evident sheath, or with a thin, soft sheath, not spiral, constricted or not constricted at the nodes, the end wall of the terminal cell thickened in some species when mature, cross walls with or without a layer of granules.

124

KEY TO THE SPECIES OF OSCILLATORIA
AS TREATED IN THE OLDER LITERATURE

1	Trichomes forming regular spirals	*O. bonnemaisonii*
1	Trichomes not forming spirals	2
2	Trichomes strictly planktonic	*O. miniata* (p. 126)
2	Trichomes not strictly planktonic	3
3	Trichomes 17–29 μm in diameter	*O. margaritifera* (p. 126)
3	Trichomes not over 12 μm in diameter	4
4	Trichomes 6–11 μm in diameter	5
4	Trichomes not over 6.5 μm in diameter	6
5	Plants epiphytic; cross walls without granules	*O. corallinae* (p. 126)
5	Plants not epiphytic; cross walls with granules	*O. nigro-viridis* (p. 127)
6	Trichomes without constrictions at the nodes	7
6	Trichomes slightly to distinctly constricted at the nodes	8
7	Cells one-third as long as wide	*O. brevis* (p. 127)
7	Cells mostly isodiametric	*O. subuliformis* (p. 127)
8	Cells longer than wide; apical cell pointed	*O. acuminata* (p. 128)
8	Cells generally isodiametric	9
9	Trichomes much constricted at the nodes	*O. salinarum* (p. 128)
9	Trichomes only slightly constricted at the nodes	10
10	Apical cell capitate	*O. amoena* (p. 128)
10	Apical cell rounded or tapering to a point	*O. laetevirens* (p. 128)

Oscillatoria bonnemaisonii Crouan. Trichomes 18–36 μm in diameter, cells 3–6 μm long, forming loose, regular spirals, the apex not tapering, nodes with slight constrictions.

125

This variant appears to be a form of *Oscillatoria margaritifera* with spirals and a form of *Lyngbya sordida* without a sheath but with spirals. It is included in *Microcoleus lyngbyaceus* by Drouet (1968). It has been recorded more often from the Pacific coast, but it is known from Bermuda, the British Isles, and elsewhere.

Gomont 1892, p. 215, pl. 6, fig. 17–18; Tilden 1910, p. 68, pl. 4, fig. 10; Geitler 1932, p. 942; Lindstedt 1943, p. 60, pl. 7, fig. 8–10; Chapman 1961, p. 21, fig. 8; Drouet 1968, fig. 108, a drawing of a trichome of the type specimen.

Oscillatoria miniata Hauck.   Trichomes 16–24 μm in diameter, cells 7–11 μm long, constricted at the nodes, the apex briefly tapering, color often dull red or maroon, especially in deeper water in which the predominant light is green or blue.

Apparently a marine planktonic form of warmer water as it is known from the Caribbean Sea, along the coast of the southeastern United States, in the Mediterranean and Adriatic seas. It is included in *Microcoleus lyngbyaceus* by Drouet (1968).

Hauck 1885, p. 236; Gomont 1892, p. 216; West 1899, p. 337; Tilden 1910, p. 68; Desikachary 1959, p. 202, pl. 40, fig. 17.

Oscillatoria margaritifera Kützing.   Trichomes large, 17–29 μm in diameter, cells 3–6 μm long, constricted at the nodes, the apex slightly tapering, olive-green, although the plant mass may be blackish green as seen on rocks in shallow water.

Known from New England to the West Indies and along the Pacific coast of North America. Also reported from Europe. This species is another synonym of *Microcoleus lyngbyaceus* (Drouet 1868).

Gomont 1892, p. 216, pl. 6, fig. 19; Tilden 1910, p. 69, pl. 4, fig. 11; Collins and Hervey 1917, p. 19; Newton 1931, p. 16, fig. 14; Fremy 1936, p. 27; 1938, p. 31; Chapman 1961, p. 22, fig. 9.

Oscillatoria corallinae Gomont.   Trichomes 6–10 μm in diameter, with constrictions at the nodes, the apex slightly tapering, cells 2.8–4.0 μm long.

This form of *Microcoleus lyngbyaceus* (Drouet 1968) is often epiphytic on larger algae. Filaments may be attached near the middle with the ends free and erect. It has been reported along both coasts of North America, from the West Indies and Hawaii. Of worldwide distribution.

Gomont 1892, p. 218, pl. 6, fig. 21; Collins 1900 p. 4; Tilden 1910, p. 70, pl. 4, fig. 16; Taylor 1928, p. 45, pl. 1, fig. 14; Newton 1931, p. 16; Fremy 1936, p. 28, fig. 6; Chapman 1961, p. 23, fig. 11.

Oscillatoria nigro-viridis Thwaites. Trichomes 7–11 μm in diameter, tapering at the tips through several cells and terminating in a somewhat conical apical cell, cells short, 3–5 μm long.

Abundant along both coasts of North America, the West Indies, Hawaii, the British Isles. Of worldwide distribution. This species is included in *Porphyrosiphon notarisii* by Drouet (1968).

Gomont 1892, p. 217, pl. 6, fig. 20; Tilden 1910, p. 69, pl. 4, fig. 12; Lindstedt 1943, p. 61, pl. 7, fig. 7; Humm and Caylor 1957, p. 234, pl. 1, fig. 8; Cocke 1967, p. 37, fig. 73.

Oscillatoria brevis Kützing. Trichomes 4.0–6.5 μm in diameter, cells 1.5–2.8 μm long, not constricted at the nodes, cross walls with granules, the apical cell long-tapering with a thin end wall and sometimes curved.

Drouet (1968) has placed this species in *Arthrospira neapolitana*. It is found in fresh and brackish water, sometimes in the sea. Newton (1931) reports the variety *neapolitana* Gomont from a marine habitat in Britain.

Kützing 1843, p. 186; Tilden 1910, p. 79, pl. 4, fig. 32; Newton 1931, p. 18; Lindstedt 1943, p. 63, pl. 7, fig. 13; Chapman 1961, p. 24, fig. 13.

Oscillatoria subuliformis Kützing. Trichomes 4.5–6.5 μm in diameter, cells mostly isodiametric, without constrictions at the nodes. Apex of trichomes tapering through several cells and terminating in a long-conical apical cell to 10 μm in length when mature.

Recorded from coastal waters of the British Isles and the entire Atlantic coast of North America; probably of worldwide distribution.

Drouet (1968) regards this name as a synonym of *Porphyrosiphon notarisii*.

Kützing 1863, p. 7: Gomont 1892, p. 226, pl. 7, fig. 10; Tilden 1910, p. 77, pl. 4, fig. 29; Newton 1931, p. 17; Fremy 1934, p. 125, pl. 31, fig. 8; Geitler 1932, p. 949, fig. 603b; Desikachary 1959, p. 213, pl. 49, fig. 10; Cocke 1967, p. 39, fig. 81.

Oscillatoria acuminata Gomont.   Trichomes 3–5 μm in diameter, cells 5–8 μm long, the apex tapering through several cells, the apical cell sharply pointed and bent, when mature, with a thin end wall.

Known from marine habitats in Italy and California and elsewhere in freshwater. This species is regarded as a form of *Schizothrix tenerrima* by Drouet (1968).

Gomont 1892, p. 227, pl. 7, fig. 12; Tilden 1910, p. 78, pl. 4, fig. 29; Desikachary 1959, p. 240, pl. 38, fig. 7; pl. 40, fig. 13; Cocke 1967, p. 45, fig. 101.

Oscillatoria salinarum Collins.   Cells isodiametric, 4 μm, the trichomes much constricted at the nodes, the apex tapering through several cells; the trichomes sometimes coiled in a regular circle.

The type material came from ditches of a salt works in Puerto Rico; it is known from several stations in the West Indies and southern Gulf of Mexico. This name is a synonym of *Schizothrix arenaria* (Drouet 1968).

Collins 1906, p. 105; Tilden 1910, p. 77; Chapman 1961, p. 22, fig. 10; Humm and Hildebrand 1962, p. 234.

Oscillatoria amoena Gomont.   Trichomes 2.5–4.2 μm in diameter, cells more or less isodiametric, with slight constrictions at the nodes, the ends tapering through several cells and terminating in an apical cell that is elongate-conical with a thickened end wall when mature.

Forming a feltlike layer or occurring as scattered trichomes in intertidal sand where wave action is low. Known from both coasts of North America and the Gulf of Mexico in brackish water. *O. amoena* is a form of *Microcoleus vaginatus* (Drouet 1968). It is much more at home in freshwater than in the sea.

Gomont 1892, p. 225, pl. 7, fig. 9; Tilden 1910, p. 77, pl. 4, fig. 26; Cocke 1967, p. 43, fig. 94.

Oscillatoria laetevirens Crouan.   Plant mass thin membranous, the trichomes held together by an invisible polysaccharide, cells mostly isodiametric, 3–5 μm, trichomes slightly constricted at the nodes, the apex tapering through several cells, often curved.

Of worldwide distribution in the intertidal zone and below, and also in fresh water. This name is a synonym of *Schizothrix arenaria* (Drouet 1968).

Crouan 1860, p. 371; Gomont 1892, p. 226, pl. 7, fig. 11; Tilden 1910, p. 78, pl. 4, fig. 28; Geitler 1932, p. 949, fig. 603c; Desikachary 1959, p. 213, pl. 39, fig. 2–3.

Lyngbya

Trichomes enclosed in a firm, distinct sheath, single within the sheath, the sheaths unbranched, sometimes lamellose, colorless or occasionally yellow-brown with age. They may or may not be constricted at the nodes and may be with or without a layer of granules against the end walls, the terminal cell in some species with a thickened end wall. Plants forming a flat, skinlike layer; a feltlike, expanded stratum, or loose skeins of long, entangled filaments, sometimes epiphytic and in tufts.

KEY TO THE SPECIES OF LYNGBYA

1   Trichomes less than 2 μm in diameter   2

1   Trichomes more than 2 μm in diameter   3

2   Trichomes spiralling around larger algae; not constricted at the nodes
L. epiphytica (p. 131)

2   Not enwrapping other algae but often within the sheath of other bluegreens; trichomes not constricted at the nodes
L. rivulariarum

3   Trichomes not over 8 μm in diameter, epiphytic   4

3   Trichomes mostly over 8 μm in diameter   5

4   Trichomes rose or purple   *L. gracilis* (p. 131)

4   Trichomes bluegreen   *L. meneghiniana* (p. 131)

5   Sheaths darkening with age   *L. aestuarii* (p. 132)

5   Sheaths remaining colorless   6

6   Trichomes constricted at the nodes, 14–31 μm in diameter
L. sordida (p. 132)

6   Trichomes not constricted at the nodes   7

7   Trichomes 5–12 μm in diameter   *L. semiplena* (p. 132)

7   Trichomes of greater diameter   8

8   Trichomes 9–25 μm in diameter, usually attached as a skin or turf
L. confervoides (p. 133)

8   Trichomes 16–60 μm in diameter, usually forming a skein or loose mat   *L. majuscula* (p. 133)

Lyngbya epiphytica Hieronymous.   Filaments spiraling around larger filamentous algae, the sheath distinct but thin and hyaline. Trichomes 1.0–1.5 μm in diameter, with slight or no contraction of the nodes; cells 1–2 μm long, the terminal cell with rounded apex and thin-walled.

This enwrapping form of *Schizothrix calcicola* is of cosmopolitan distribution.

130

Fremy 1929, p. 165, fig. 162; 1934, p. 100, pl. 29, fig. 4; Humm and Caylor 1957, p. 235, pl. 2, fig. 3; Desikachary 1959, p. 284, pl. 53, fig. 7; Cocke 1967, p. 56, fig. 123; Drouet 1968, fig. 11, drawing of a trichome from the type specimen.

Lyngbya rivulariarum Gomont.   Filaments growing within the sheath of larger bluegreen algae, their own sheaths thin, colorless, trichomes 0.7–0.8 μm in diameter, slightly constricted at the nodes, the cells 2.3 μm long, without granules on the cross walls, the tip cell rotund, the end wall thin.

This form of *Schizothrix calcicola* usually is found within the sheath of larger bluegreens, but not always.

Gomont 1892, p. 148; Tilden 1910, p. 111; Newton 1931, p. 25; Geitler 1932, p. 1048; Fremy 1934, p. 112, pl. 29, fig. 6; Desikachary 1959, p. 293.

Lyngbya gracilis (Meneghini) Rabenhorst.   Trichomes 5–8 μm in diameter; the cells about half as long as wide, some constriction at the cross walls, the apex not tapering, and the terminal cell with a rounded end and thin-walled, the cell contents rose-colored, and the plant mass maroon to purple.

This form of *Schizothrix mexicana, sensu* Drouet (1968) is usually an in-conspicuous epiphyte on larger algae and is apparently inseparable from *L. meneghiniana.* Cosmopolitan.

Rabenhorst 1865, p. 145; Gomont 1892, p. 124, pl. 2, fig. 20; Tilden 1910, p. 117, pl. 5, fig. 36; Howe 1920, p. 622; Fremy 1934, p. 102, pl. 26, fig. 3; 1938, p. 25; Humm and Darnell 1959, p. 270; Desikachary 1959, p. 285, pl. 52, fig. 2; Chapman 1961, p. 27, fig. 19.

Lyngbya meneghiniana (Kützing) Falkenberg.   The description of this form of *Schizothrix mexicana* as given by Tilden (1910) is essentially the same as the above except that the cell contents are said to be bluegreen and the apical cell has a slightly thickened outer wall.

Reported from the Atlantic coast of North America, the British Isles, the Adriatic Sea, and Hawaii.

Gomont 1892, p. 125; Tilden 1910, p. 117; Newton 1931, p. 23; Chapman 1961, p. 30, fig. 24; Cocke 1967, p. 60, fig. 138.

Lyngbya aestuarii Gomont.   Trichomes 8–24 μm in diameter, without constrictions at the nodes; the apex slightly tapering, the cells 2.6–5.6 μm long. Plants often abundant on soft, unconsolidated sediments in bays and estuaries, sometimes breaking loose and drifting as blackish-green mats, often floating as a result of the entrapment of oxygen bubbles; common in salt marshes and mangrove swamps.

There are many records of this form of *Microcoleus lyngbyaceus sensu* Drouet (1968) along the Atlantic coast of North America from Prince Edward Island to the Caribbean Sea, along most of the Pacific coast of North America, and along the coast of Europe including the Mediterranean Sea. Of worldwide distribution.

Gomont 1892, p. 127, pl. 3, fig. 1–2; Tilden 1910, p. 120, pl. 5, fig. 40–41; Fremy 1929, p. 183, fig. 152; 1934, p. 104, pl. 27, fig. 1–5; Newton 1931, p. 23, fig. 17; Humm and Jackson 1955, p. 241; Humm and Caylor 1957, p. 236, pl. 2, fig. 5; Almodovar and Blomquist 1959, p. 167; Chapman 1961, p. 27, fig. 20; Cocke 1967, p. 58. fig. 131.

Lyngbya sordida (Zanardini) Gomont.   Trichomes 14–31 μm in diameter, the cells short, 4–10 μm long; the apical cell hemispherical, without a thickened end wall, trichomes not tapering at the tips, constricted at the nodes; the cytoplasm with scattered granules but these not concentrated at the end walls.

This species is usually an epiphyte on other algae and seagrasses. It is one of the largest ecophenes of *Schizothrix mexicana* (Drouet 1968).

Gomont 1892, p. 126, pl. 2, fig. 21; Tilden 1910, p. 118, pl. 5, fig. 37; Taylor 1928, p. 45, pl. 1, fig. 22 (as *L. rosea* Taylor); Chapman 1961, p. 26, fig. 18.

Lyngbya semiplena (C. Agardh) J. Agardh.   Trichomes 5–12 μm in diameter, the cells 2–3 μm long; nodes not constricted, apex of trichomes slightly tapering, conspicuous granules on the cross walls; plant mass often turf-forming on solid substrata; also commonly epiphytic.

Of worldwide distribution. This species is another synonym of *Microcoleus lyngbyaceus* (Drouet 1978).

Gomont 1892, p. 138, pl. 3, fig. 7–11; Tilden 1910, p. 118, pl. 5, fig. 38; Geitler 1932, p. 1061, fig. 672a; Fremy 1934, pl. 108, pl. 28, fig. 3; Lindstedt 1943, p. 83, pl. 5, fig. 5–6, 9; Humm and Caylor 1957, p. 236, pl. 2, fig. 4; Desikachary 1959, p. 315, pl. 49, fig. 8; pl. 52, fig. 7; Cocke 1967, p. 60, fig. 139.

Lyngbya confervoides C. Agardh.   Trichomes 9–25 μm in diameter, the cells 2–4 μm long, not constricted at the nodes, the apex not tapering.

This is one of the most common forms of *Microcoleus lyngbyaceus*. On some substrata it produces tufts with the filaments in fascicles up to several centimeters in height. On rocks and pilings it often forms a flat skin that can be peeled off. It is the form most resistant to the toxic compounds of creosoted pilings on which it often forms a pure stand.

It occurs along the entire coastline of North America and is of worldwide distribution.

Gomont 1892, p. 136, pl. 3, fig. 5–6; Tilden 1910, p. 119, pl. 5, fig. 39; Taylor 1928, p. 44, pl. 1, fig. 20; Fremy 1936, p. 24; 1938, p. 29; Humm and Caylor 1957, p. 235, pl. 2, fig. 1; Joly 1957, p. 173, pl. 12, fig. 11; Chapman 1961, p. 28, fig. 22a–b; Cocke 1967, p. 61, fig. 140.

Lyngbya majuscula Gomont.   Trichomes 16–20 μm in diameter, the cells only 2–4 μm long, without constrictions at the nodes, the apex not tapering. Plants forming blackish-green or olive-brown mats on the bottom in quiet, shallow water. These mats often tear loose and float, buoyed up by entrapped oxygen bubbles, and may drift for long distances. They sometimes resemble a toupe.

This is the largest form of *Microcoleus lyngbyaceus* in trichome diameter and in trichome length and plant mass as well. It is essentially tropical or subtropical, although there are numerous records from New England and farther north along the Atlantic coast, especially during summer. Loose, floating plants are carried north in the Gulf Stream and may be blown out of the stream north of Cape Hatteras. It is often entangled in pelagic *Sargassum*.

Gomont 1892, p. 131, pl. 3, fig. 3–4; Tilden 1910, p. 123, pl. 5, fig. 42; Taylor 1928, p. 44, pl. 1, fig. 19; Geitler 1932, p. 1060, fig. 672c, d; Fremy 1934, p. 106, pl. 28, fig. 1; 1936, p. 24; Lindstedt 1943, p. 84, pl. 10, fig. 11; Blomquist and Humm 1946, p. 3, pl. 1, fig. 5; Chapman 1956, p. 357, fig. 4, no. 9; Desikachary 1959, p. 313, pl. 48, fig. 7; Cocke 1967, p. 61, fig. 141.

Phormidium

Plants forming a feltlike layer or a thin, flat stratum; the sheaths thin and partly agglutinated or entirely diffluent. Sheaths, where distinct, not branched. Trichomes cylindrical or constricted at the nodes in some species, the tips often tapering; the end wall of the terminal cell thickened in some.

KEY TO THE SPECIES OF PHORMIDIUM

1   Trichomes not over 3 μm in diameter   **2**

1   Trichomes over 3 μm in diameter   **5**

2   Trichomes embedded in a firm, gelatinous, hemispherical or irregular mass   *P. crosbyanum*

2   Trichomes forming a thin, flat layer   **3**

3   Cells mostly isodiametric   *P. fragile* (p. 136)

3   Cells mostly longer than the diameter   **4**

4   Trichomes pink or rose-colored   *P. persicinum* (p. 136)

4   Trichomes green or blue-green   *P. subuliforme* (p. 137)

5   Trichomes without constrictions at the nodes   **6**

5   Trichomes with constrictions at the nodes   **7**

6   Plants restricted to sponges, ending in tufts or streamers   *P. spongeliae* (p. 137)

6   Plants in a flat, thin layer, not (or rarely) on sponges   *P. autumnale* (p. 137)

7   Plants forming erect tufts; trichomes 6–9 μm in diameter   *P. penicillatum* (p. 137)

7   Plants in a flat, thin layer, trichomes 5 μm in diameter   *P. submembranaceum* (p. 138)

Phormidium crosbyanum Tilden. Plants forming firm, gelatinous masses to 5 cm in diameter or more and about 2 cm thick, hemispherical to irregular, attached to rocks in shallow water of high salinity, blackish green in color, and resembling the small chicken liver sponge of south Florida and the West Indies. Trichomes embedded in the fused, cartilaginous extracellular polysaccharide, 1–2 μm in diameter; the cells 1–3 μm long, constricted at the nodes, the apical cell acute-conical.

This is an ecophene of shallow tropical waters of high salinity that was first described by Dr. Josephine Tilden from the island of Oahu, Hawaii. Her statement "sheaths extremely delicate" apparently refers to a thin lamina of extracellular polysaccharide that is distinguishable immediately around the trichomes. Her statement "impregnated with lime" is puzzling, as plants from the Florida Keys and West Indies seem to have very little or no calcium carbonate embedded within them.

This form of *Schizothrix calcicola* is convenient material for study of the chemical nature and physical properties of the extracellular polysaccharides from plants growing in the natural environment as it can be obtained in quantity.

Tilden 1910, p. 96, pl. 4, fig. 60–61; Drouet 1942, p. 119; Chapman 1961, p. 25, fig. 16; Humm 1963, p. 517.

Phormidium fragile (Meneghini) Gomont.   Sheaths diffluent to form a gelatinous plant mass as a thin stratum, although recently produced extracellular polysaccharide may be discernable as a sheath around the trichomes. Cells 1.2–2.3 μm in diameter, with constrictions at the nodes; the terminal cell acute-conical.

This form of *Schizothrix calcicola* often produces thin layers on or inside submerged bottles, on walls of marine aquaria, and various substrata. It is abundant from New England to the West Indies, around the Gulf of Mexico, along the Pacific coast of North America, Canary Islands, coastal waters of Britain. Cosmopolitan.

Gomont 1892, p. 163, pl. 4, fig. 13–15; Tilden 1910, p. 93, pl. 4, fig. 52–53; Collins and Hervey 1917, p. 20; Newton 1931, p. 19; Fremy 1934, p. 86, pl. 22, fig. 6; Lindstedt 1943, p. 65, pl. 8, fig. 3–4; Desikachary 1959, p. 253, pl. 44, fig. 1–3; Cocke 1967, p. 48, fig. 104.

Phormidium persicinum (Reinke) Gomont.   Filments forming delicate masses or a thin, papery layer on shells, stones, other algae, usually rather loosely entangled, the color bluegreen in shallow water, reddish in deeper water or in low light intensity. Sheaths diffluent. Trichomes 1.7–2.0 μm in diameter, the cells 2–7 μm long, slightly constricted at the nodes. The terminal cell acute-conical.

Common along the Atlantic coast of North America from New England to the Caribbean Sea, and along the Pacific coast; of world wide distribution but not often reported. This species is a form of *Schizothrix calcicola* according to Drouet (1968).

Gomont 1892, p. 164; Tilden 1910, p. 94; Taylor 1928, p. 46, pl. 1, fig. 16; Fremy 1934, p. 86, pl. 22, fig. 7.

Phormidium subuliforme Gomont.   Trichomes 2.0–2.8 μm in diameter, cells 6–8 μm long, apex of trichome tapering, often bent, slightly constricted at the nodes, the apical cell acute-conical with a thin end wall.

Of worldwide distribution, usually in the intertidal zone or a little below on shells, stones, submerged bottles. According to Drouet (1968) this plant is a form of *Schizothrix arenaria*.

Gomont 1892, p. 169, pl. 4, fig. 26; Tilden 1910, p. 99, pl. 4, fig. 67; Humm and Hildebrand 1962, p. 235.

Phormidium spongeliae Gomont.   Trichomes 7.5–8.5 (up to 12?) μm in diameter, cells about one half as long to equal in diameter and length, not constricted at the cross walls, the terminal cell with rounded end.

This species has been reported only from the canals of sponges. It has been found coating the inner walls, with filaments streaming out of the osculum. Probably common in tropical waters. It is a synonym of *Schizothrix mexicana*.

Gomont 1892, p. 161, pl. 4, fig. 8–10; Feldmann 1933, p. 384, fig. 1–2; Fremy 1934, p. 84, pl. 22, fig. 3–4.

Phormidium autumnale (Agardh) Trevisan.   Trichomes 4–7 μm in diameter, cells short, 2–5 μm long and not constricted at the nodes. Sheaths sometimes distinct though usually diffluent and forming a thin layer on the substrate. Apex of trichome briefly tapering, the conical apical cell with a distinct thickening of the end wall.

Known from many brackish water environments along both Atlantic and Pacific coasts, especially in temperate and cooler areas. *P. autumnale* is a synonym of *Microcoleus vaginatus* (Drouet 1968).

Gomont 1892, p. 187, pl. 5, fig. 23–24; Tilden 1910, p. 107, pl. 5, fig. 18–19;Newton 1931, p. 20; Fremy 1934, p. 93, fig. 24; Lindstedt 1943, p. 8, fig. 15–16; Cocke 1967, p. 53, fig. 121.

Phormidium penicillatum Gomont.   Plants usually chestnut brown, in brushlike clusters attached at the base, trichomes with slight constrictions at the nodes, 6 μm in diameter, in delicate sheaths that are soon diffluent; cells twice as long as the diameter, 7–12 μm long, the terminal cell with a thickened end wall when mature.

Phormidium penicillatum is a warm water marine form of *Oscillatoria submembranacea* (Drouet 1968). It has been recorded from the West Indies and Indian Ocean.

Gomont 1893, p. 88; Chapman 1961, p. 25, fig. 17. Chapman's fig. 17 is taken from fig. 18 of Taylor 1928, an illustration of *Symploca profunda* Taylor, which is not synonymous with *Phormidium penicillatum.* The former is a synonym of *Schizothrix mexicana* (Drouet 1968).

Phormidium submembranaceum (Ardissone and Strafforello) Gomont.   Plant mass a thin or leathery layer of diffluent sheaths with trichomes about 5 μm in diameter having nodal constrictions, the apex tapering through a few cells, cells 4–10 μm long.

Recorded from the Atlantic and Pacific coasts of North America and from the coasts of Europe and India. *P. submembranaceum* is a synonym of *Oscillatoria submembranacea* Ardissone and Strafforello (Drouet 1968).

Gomont 1892, p. 180, pl. 5, fig. 13; Tilden 1910, p. 104, pl. 5, fig. 7–8; Geitler 1932, p. 1023, fig. 652; Fremy 1934, p. 91, pl. 24, fig. 2; Desikachary 1959, p. 273, pl. 44, fig. 19; Cocke 1967, p. 52, fig. 118.

Symploca

Plants forming erect, wicklike bundles that arise from prostrate filaments, sheaths often branched, Trichomes single within a sheath, the sheath distinct, firm, colorless, the trichomes constricted or not constricted at the nodes, the terminal cell with or without a thickened end wall.

KEY TO THE SPECIES OF SYMPLOCA

1   Trichomes 1.5–3.5 μm in diameter, much constricted at the nodes, the fascicles appressed   *S. laete-viridis*

1   Trichomes of greater diameter   2

2   Trichomes 4–6 μm in diameter, constricted at nodes from base to apex   *S. atlantica*

2   Trichomes 6–14 μm in diameter, constricted at nodes near apices only   *S. hydnoides*

Symploca laete-viridis Gomont.   Trichomes 1.5–3.0 μm in diameter, cells about twice as long as wide, much constricted at the nodes, filaments in groups forming tufts or fascicles about 1 mm high on rocks or other solid substrata.

This form of *Schizothrix arenaria* has been recorded from Key West, Florida, Alaska, and India.

Gomont 1892, p. 109, pl. 2, fig. 6–8; Setchell and Gardner 1903, p. 188; Tilden 1910, p. 130, pl. 5, fig. 50; Geitler 1932, p. 1121, fig. 728; Desikachary 1959, p. 335.

Symploca atlantica Gomont.   Plants producing wicklike bundles to a height of 1 cm, sheaths distinct and containing a single trichome; false branches occasional, trichomes 4–6 μm in diameter, constricted at the nodes, cells 2–6 μm long, the apex of the trichome tapering through several cells; apical cell with thickened end wall.

Symploca atlantica is considered by Drouet (1968) to be a synonym of *Oscillatoria submembranacea*.

Gomont 1892, p. 109, pl. 2, fig. 5; Tilden 1910, p. 129, pl. 5, fig. 48; Chapman 1961, p. 32, fig. 29; Cocke 1967, p. 65, fig. 148; Drouet 1968, fig. 62, a drawing of a trichome from the type specimen.

Symploca hydnoides Kützing.   Plants forming a turf of erect, wicklike bundles, some with false branching; trichomes 6–14 μm in diameter, cells mostly isodiametric, slightly constricted at the nodes, the apical cell often slightly inflated and rounded, with a thin end wall.

Symploca hydnoides is a synonym of *Calothrix hydnoides* Harvey, one of Drouet's *nomina excludenda* (1968). However, the name has been interpreted differently, especially as varieties and forms. These are synonyms of *Schizothrix mexicana* except for a small plant referred to as *forma minor* Iyengar and Desikachary which is a synonym of *Schizothrix calcicola*.

Kützing 1849, p. 272; Gomont 1892, p. 106, pl. 2, fig. 1–4; Tilden 1910, p. 129, pl. 5, fig. 49; Lindstedt 1943, p. 74, pl. 9, fig. 3–4; Chapman 1961, p. 31, fig. 27.

Plectonema

Plants with branched sheaths ("false branching"); branches single or in pairs; filaments entangled among other small algae or the plants producing feltlike masses or small, dense balls; sheaths usually thin and firm, colorless or yellowish; trichomes usually constricted at the nodes.

KEY TO THE SPECIES OF PLECTONEMA

1   Trichomes penetrating shells or other forms of limestone
                                                        P. terebrans

1   Trichomes not penetrating limestone   **2**

2   Trichomes 1–1.5 μm in diameter   *P. nostocorum*

2   Trichomes 1–4 μm in diameter   **3**

3   Trichomes long and entangled, forming dense balls
                                        P. calothrichoides (p. 142)

3   Trichomes forming a thin, flat layer   *P. battersii* (p. 142)

Plectonema terebrans Gomont.   Penetrating all forms of limestone in shallow water, especially living and dead mollusk shells, living and dead stony corals, calcareous annelid tubes, and bryozoan tests. Trichomes 0.9–1.5 μm in diameter, cells 2–6 μm long, without constrictions at the nodes, the cross walls often with two granules on each side, and the terminal cell with rounded end; false branches often present.

This form of *Schizothrix calcicola* is widely distributed the entire length of both coasts of North America as well as British coastal water. The surface layers of limestone must be decalcified with dilute HCL to see these plants.

Gomont 1892, p. 103; Tilden 1910, p. 209, pl. 11, fig. 6; Newton 1931, p. 25; Desikachary 1959, p. 435, pl. 61, fig. 4–5; Cocke 1967, p. 123, fig. 255.

Plectonema nostocorum Bornet.   Filaments with false branches that are solitary or in pairs; the sheath thin and hyaline, trichomes 1.0–1.5 μm in diameter, cells 2.0–2.5 μm long, and the nodes somewhat constricted, the apical cell rounded with a thin end wall.

Recorded from New England to the West Indies, the British Isles, and Hawaii. It sometimes forms a loosely cementing layer or mat in the sand around mangrove islands in tropical waters in association with the *Schizothrix longiarticulata* form of *Schizothrix arenaria*. This slender *Plectonema* is a synonym of *Schizothrix calcicola* (Drouet 1968).

Gomont 1892, p. 102, pl. 1, fig. 11; Tilden 1910, p. 209, pl. 11, fig. 7; Lindstedt 1943, p. 77, pl. 9, fig. 8; Desikachary 1959, p. 439, pl. 83, fig. 7; Cocke 1967, p. 125, fig. 260.

Plectonema calothrichoides Gomont.   Filaments entangled in dense balls and tending to be radial in arrangement, tapering at the apices, having false branches in pairs, the trichomes 2.0–2.5 μm in diameter, the cells shorter than their diameter with nodes constriction, apical cell rounded at the end.

A form of *Schizothrix calcicola,* known from New England to Florida. Cosmopolitan.

Gomont 1899, p. 30, pl. 1, fig. 6–10; Tilden 1910, p. 210, pl. 11, fig. 10; Geitler 1932, p. 683, fig. 437b; Fremy 1934, p. 96, pl. 25; Lindstedt 1943, p. 76, pl. 9, fig. 7; Cocke 1967, p. 123, fig. 256.

Plectonema battersii Gomont.   Filaments with false branches, often in pairs; sheaths hyaline, thin to moderately thick. Trichomes 3.0–3.5 μm in diameter, the cells 3–4 times shorter than the diameter, with constrictions at the nodes, the tips of the trichomes tapering, the apical cell rounded.

Often epiphytic on the Coralline algae. It develops extensive mats that loosely cement the sand of gently sloping open intertidal zones. Like *P. nostocorum,* this species is another synonym of *Schizothrix calcicola* (Drouet 1968).

Gomont 1899, p. 36; Tilden 1910, p. 211; Newton 1931, p. 25, fig. 18; Cocke 1967, p. 124, fig. 257a–b.

Hydrocoleum

Plants forming an expanded, caespitose cushion or, sometimes, a flat layer, trichomes one to several within a sheath, the sheaths occasionally branched or becoming soft and diffluent, cells generally shorter than the diameter of the trichomes, the terminal cell with a thickened end wall.

KEY TO THE SPECIES OF HYDROCOLEUM

1 Trichomes constricted at the nodes, 14–21 μm in diameter
                                                         H. comoides

1 Trichomes not constricted at the nodes   2

2 Trichomes 8–16 μm in diameter   *H. lyngbyaceum*

2 Trichomes 14–24 μm in diameter   3

3 Plant mass caespitose, turflike, to 2 cm tall   *H. cantharidosmum*
                                                         (p. 144)

3 Plants not caespitose, forming a flat, gelatinous growth
                                              H. glutinosum (p. 144)

Hydrocoleum comoides Gomont.   Trichomes 14–21 μm in diameter, the cells 3–5 μm long, with constrictions at the nodes and the apex usually tapering through several cells. Plants usually in the form of a cushion-shaped group of erect filaments with occasional false branches.

Apparently a warm water form of *Microcoleus lyngbyaceus* of worldwide distribution. The type came from Australia.

Gomont 1892, p. 335, pl. 12, fig. 3–5; Tilden 1910, p. 134, pl. 5, fig. 56; Collins and Hervey 1917, p. 23; Howe 1920, p. 623; Williams 1948, p. 684; Drouet 1942, p. 71; Cocke 1967, p. 79, fig. 175a–b.

Hydrocoleum lyngbyaceum Kützing.   Trichomes 8–16 μm in diameter, the cells 2.5–4.5 μm long, not constricted at the nodes, the apex of the trichome tapering. Filaments with false branching, one to several trichomes within a sheath; granules on the cross walls.

This form of *Microcoleus lyngbyaceus* is abundant and widely distributed.

Kützing 1849, p. 259; Gomont 1892, p. 337, pl. 12, fig. 8–10; Tilden 1910, p. 135, pl. 5, fig. 58 (as *Hydrocoleus lyngbyaceus*); Taylor 1928, p. 43, pl. 1, fig. 17; Newton 1931, p. 29, fig. 20; Fremy 1929, p. 89, fig. 88; 1934, p. 72, pl. 19, fig. 1; Lindstedt 1943, p. 91, pl. 11, figs. 6, 10; Humm and Caylor 1957, p. 238, pl. 3, fig. 1–2; Chapman 1961, p. 37, fig. 36a–b; Seoane-Camba 1965, p. 51, fig. 17, no. 4.

Hydrocoleum cantharidosmum (Montagne) Gomont. Trichomes 18–24 μm in diameter, the cells 2–4 μm long, not constricted at the nodes, the apex tapering through several cells, trichomes single within the sheath in the upper, younger parts of a plant mass, often several within a sheath in the lower, older parts. Plants usually form tufts to 2 cm tall on stones or shells or on larger algae.

This is a warm water form of *Microcoleus lyngbyaceus* that was described from the Canary Islands and has been reported from Hawaii, the West Indies, and Florida.

Gomont 1892, p. 336, pl. 12, fig. 6–7; Vickers 1905, p. 95; Tilden 1910, p. 135, pl. 5, fig. 57 (as *Hydrocoleus cantharidosmus*); Howe 1920, p. 623; Fremy 1938, p. 19.

Hydrocoleum glutinosum Gomont. Trichomes 14–21 μm in diameter, cells 2.5–3.5 μm long, not constricted at the nodes, tapering at the apex through several cells, the terminal cell with a thickened end wall when mature, the cross walls with granules. Plant mass forming a flat layer, not tufted, the sheaths soft and somewhat diffluent; epiphytic or on stones and shells.

This form of *Microcoleus lyngbyaceus* is cosmopolitan.

Gomont 1892, p. 339; Tilden 1910, p. 136, pl. 5, fig. 59; Fremy 1934, p. 73, pl. 19, fig. 2; Lindstedt 1943, p. 92, pl. 11. fig. 7; Chapman 1961, p. 36, fig. 35a–b.

Microcoleus

Trichomes numerous within the sheath in well-developed filaments, densely crowded and often twisted, forming horizontal masses of entangled filaments or mats, the sheaths occasionally branched; apical cell elongate-conical, the end wall thin (in species included here).

KEY TO THE SPECIES OF MICROCOLEUS

Trichomes 1.5–2.0 μm in diameter   *M. tenerrimus*

Trichomes 2.5–6.0 μm in diameter   *M. chthonoplastes*

Microcoleus tenerrimus Gomont.   Trichomes 1.5–2.0 μm in diameter, constricted at the nodes and without granules on the cross walls, the cells 2–6 μm long; trichomes few to many within a distinct sheath, especially in the intertidal zone, or the sheath somewhat diffluent. Apical cell, when mature, long acute-conical, the end wall thin.

Cosmopolitan in temperate and tropical regions in damp places, in freshwater, and in the intertidal zone of marine habitats. It is a synonym of *Schizothrix tenerrima* (Drouet 1968).

Gomont 1892, p. 355, pl. 14, fig. 9–11; Tilden 1910, p. 155, pl. 6, fig. 27; Fremy 1934, p. 68, fig. 17; Lindstedt 1943, p. 90, pl. 11, fig. 8; Humm and Caylor 1957, p. 238, pl. 2, fig. 9; Cocke 1967, p. 75, fig. 167a–b.

Microcoleus chthonoplastes Thuret.   Trichomes 2.5–6.0 μm in diameter, cells 3.6–10.0 μm long, the apical cell elongate-conical with a thin end wall, many trichomes in a distinct sheath and often twisted, especially in the intertidal zone, where the plants form a thin layer or mat in protected places; filaments horizontal and intertwined, tending to stabilize the substrate.

This form of *Schizothrix arenaria* is cosmopolitan from cold waters to the tropics in the sea and also in freshwater and on damp soil.

Gomont 1892, p. 353, pl. 14, fig. 5–8; Tilden 1910, p. 155, pl. 6, fig. 28; Newton 1931, p. 27, fig. 10; Fremy 1929, p. 78, fig. 78; 1934, p. 67, pl. 17, fig. 7; Lindstedt 1943, p. 89, pl. 11, fig. 5, 9, 11; Humm and Caylor 1957, p. 238, pl. 2, fig. 10; Chapman 1961, p. 35, fig. 33a–b; Cocke 1967, p. 76, fig. 170.

Schizothrix

Trichomes single or few to numerous within a sheath, the sheaths frequently branched, filaments may form erect *Symploca*-like fascicles or a flat stratum on damp soil, in fresh water or in the sea.

KEY TO THE SPECIES OF SCHIZOTHRIX

Trichomes 1.0–1.5 µm in diameter	*S. lacustris*
Trichomes 1.7–2.0 µm in diameter	*S. longiarticulata*

Schizothrix lacustris A. Braun.   Colonies cushion-shaped or crustose, filaments crowded and repeatedly false-branched; trichomes 1.0–1.5 µm in diameter, one to many in a sheath, often spirally twisted; cells longer than the diameter, to 4 µm.

This form of *Schizothrix calcicola* is widely distributed in freshwater and has been reported in marine habitats in a few places.

A. Braun in Kützing 1849, p. 320; Gomont 1892, p. 301, pl. 6, fig. 9–12; Fremy 1929, p. 96, fig. 92; Geitler 1932, p. 1092, fig. 698–689; Fremy 1934, p. 77, pl. 20, fig. 4; Chapman 1956, p. 359, fig. 4, no. 6; Desikachary 1959, p. 325, pl. 56, fig. 6, 10.

Schizothrix longiarticulata Gardner.   Trichomes 1.7–2.0 µm in diameter, without constrictions at the nodes, one or two in a sheath, occasional false branching; cells 8.0–12.5 µm long, the apical cell conical with a thin end wall.

This form of *Schizothrix arenaria* with unusually long cells in relation to the diameter was described from Puerto Rico. It has been found in intertidal sand in Florida and mixed with other bluegreens on rocks.

Gardner 1927, p. 50, pl. 10, fig. 95 (as *Hypheothrix longiarticulata*); Nielsen 1954, p. 32; Chapman 1961, p. 31, fig. 26.

Amphithrix

Plants crustaceous, in tufts or forming a firm turf; of two layers, the lower consisting of densely interwoven filaments or torulose chains of cells; the upper consisting of erect, closely-packed filaments that taper to fine points; without heterocysts.

The two species listed below are considered to be synonymous with *Schizothrix calcicola* by Drouet (1968).

146

KEY TO THE SPECIES OF AMPHITHRIX

Cells mostly isodiametric   *A. janthina*

Cells shorter than their diameter   *A. violacea*

Amphithrix janthina (Montagne) Bornet and Flahault. Plants crustaceous, 0.5–5.0 mm thick; the filaments 1.5–2.2 in diameter, sheaths thin, the cells isodiametric and pale bluegreen.

Known from New England to North Carolina in fresh- and saltwater.

Bornet and Flahault 1886–1888 (1886), p. 344; Tilden 1910, p. 253, pl. 16, fig. 3; Cocke 1967, p. 150, fig. 295.

Amphithrix violacea (Kützing) Bornet and Flahault.   Plants in tufts or forming a turf; on rocks or epiphytic, the filaments partially horizontal and bent upward, the erect portion tapering to a fine point; 2–3 μm in diameter, 1–3 mm in height; trichomes constricted at the nodes, the cells shorter than their diameter.

Known from both freshwater and the sea; in marine habitats it is usually intertidal or in the splash zone.

Kützing 1849, p. 344; Bornet and Flahault 1886–1888 (1886), p. 344; Collins 1900, p. 41; Tilden 1910, p. 253, pl. 16, fig. 4; Newton 1931, p. 32; Geitler 1932, p. 573.

Calothrix

The genus *Calothrix* was defined in the older literature as follows: Trichomes tapering from the base upward and terminating in a multicellular, colorless hair; the sheaths distinct, branched or unbranched, the false branches separate from the filaments from which they arose; heterocysts basal only or basal and intercalary; spores, if present, basal; plants generally in the form of penicillate tufts or forming a short turflike expansion.

All species of *Calothrix* listed below are considered to be synonyms of *Calothrix crustacea sensu* Drouet 1973 with the exception of *Calothrix pilosa*. The latter species does not taper to a hair at the tips and is a synonym of *Scytonema hofmannii* (Drouet 1973).

147

KEY TO THE SPECIES OF CALOTHRIX

The following key includes only those species of *Calothrix*, as treated in the older literature (prior to Drouet 1973), that have been reported repeatedly from a marine habitat.

1   Plants within the polysaccharide of the red alga, *Nemalion*
                                         C. parasitica (p. 149)

1   Plants not endophytic, or only partly so   2

2   Filaments with or without false branches, the false branches separate, free   3

2   Filaments with false branches that remain within the original sheath, or all trichomes of the plant inside a common gelatinous matrix   13

3   Heterocysts basal only   4

3   Heterocysts basal and intercalary   9

4   Plants epiphytic   5

4   Plants on rocks, wood; rarely epiphytic   7

5   Trichomes 7–8 μm in diameter, violet   *C. fusco-violacea* (p. 149)

5   Trichomes of greater diameter   6

6   Trichomes 12 μm in diameter, to 0.5 mm tall   *C. consociata* (p. 149)

6   Trichomes 10–18 μm in diameter, 2–3 mm tall
                                        C. confervicola (p. 150)

7   Trichomes 6–8 μm in diameter   *C. contarenii* (p. 150)

7   Trichomes of greater diameter   8

8   Trichomes 8–12 μm in diameter, 2–3 mm tall   *C. pulvinata* (p. 150)

8   Trichomes 8–15 μm in diameter, to 1 mm tall
                                        C. scopulorum (p. 151)

9   Trichomes 9–12 μm in diameter, to 0.5 mm tall
                                        C. aeruginea (p. 151)

9   Trichomes generally wider and taller   10

10   Plants often epiphytic, without false branches; trichomes 8–15 μm in diameter, 1–2 mm tall   *C. crustacea* (p. 151)

10   Plants not epiphytic or rarely so, 2–5 mm tall   11

11   Trichomes 8–12 μm in diameter, 2 mm tall; false branches solitary   *C. prolifera* (p. 151)

11   False branches in groups of two or more   12

12   False branches in pairs; trichomes 9–15 μm in diameter, 3–5 mm tall   *C. vivipara* (p. 152)

12   False branches often in fasciculate groups   13

13   Trichomes ending in a hair   *C. fasciculata* (p. 152)

13   Trichomes tapering briefly at apex, not ending in a hair, the apical cell hemispherical   *C. pilosa* (p. 152)

Calothrix parasitica (Chauvin) Thuret.   Plants growing in the extracellular polysaccharide of larger algae; solitary or in small groups; trichomes 3–8 μm in diameter, tapering abruptly to a fine hair, with a basal heterocyst; cells variable in length but mostly about one-third as long as wide, with slight constrictions at the nodes.

Recorded in soft-gelatinous species of Rhodophyceae, Phaeophyceae, and Myxophyceae. Cosmopolitan.

Bornet and Flahault 1886, p. 357; Tilden 1910, p. 260, pl. 16, fig. 15–16; Chapman 1961, p. 45, fig. 47.

Calothrix fusco-violacea Crouan.   Plants forming rounded, velvety patches that are usually violet in color on larger algae. Trichomes 7–8 μm in diameter constricted at the nodes, cells shorter than their diameter; heterocysts basal. Sheaths thin and colorless.

Bornet and Flahault 1886, p. 352; Collins 1900, p. 41; Tilden 1910, p. 258, pl. 16, fig. 10.

Calothrix consociata (Kützing) Bornet and Flahault.   Plants epiphytic, often in fasciculate groups, the base curved and attached to the host below the curve; trichomes 12 μm in diameter, cells about one-third as long as the diameter, olive-green; heterocysts basal.

Bornet and Flahault 1886, p. 351; Tilden 1910, p. 257, pl. 16, fig. 9; Newton 1931, p. 33.

Calothrix confervicola (Roth) C. Agardh.   Plants epiphytic on larger algae in gregarious and fasciculate groups to 3 mm high, 12–25 μm wide; slightly or not at all enlarged at base; plant mass dark green but drying to purple or violet; trichomes 10–20 μm in diameter at bases, gradually tapering to a fine hair; terminal cells about 3–4 times shorter than the diameter; one or two basal heterocysts.

Widely distributed in the sea. Cosmopolitan.

C. Agardh 1824, p. 70; Tilden 1910, p. 256, pl. 16, figs. 6–8; Howe 1920, p. 625 (as *C. parasitica* Thuret); Taylor 1928, p. 50, pl. 2, fig. 7; Geitler 1932, p. 601, fig. 376; Fremy 1934, p. 140, pl. 35, fig. 1; 1936, p. 32; 1938, p. 34; Blomquist and Humm 1946, p. 3, pl. 1, fig. 7; Fan 1956, p. 169, fig. 6; Chapman 1956, p. 366, fig. 8, no. 2; 1961, p. 43, fig. 44; Cocke 1967, p. 156, fig. 302.

Calothrix contarenii (Zanardini) Bornet and Flahault.   Plant mass crustlike, firm, dark green; filaments closely packed, parallel, up to 1 mm long, and 9–15 μm wide; swollen at the base, sheaths fairly thick and colorless to yellow-brown; trichomes 6–8 μm in diameter, terminating in a long hair; cells as long as wide or shorter, heterocysts basal.

Attaches to submerged objects. Cosmopolitan from Arctic to tropics.

Bornet and Flahault 1886, p. 355; Collins 1891, p. 336; Tilden 1910, p. 259, p. 16, fig. 13; Newton 1931, p. 33; Geitler 1932, p. 600; Fremy 1934, p. 142, pl. 3b, fig. 4; Desikachary 1959, p. 524, pl. 111, fig. 2, 5–8.

Calothrix pulvinata (Mertens) C. Agardh.   Plants forming a spongy, expanded turf, especially on submerged wood; dull green, 2–3 mm high; trichomes 8–12 μm in diameter, tapering to a short hair tip, cells one-half to one-third as long as their diameter; sheaths thick, firm, false branches occasional, often opposite; heterocysts basal only.

C. Agardh 1824, p. 71; Bornet and Flahault 1886, p. 356; Tilden 1910, p. 260, pl. 16, fig. 14; Newton 1931, p. 33; Geitler 1932, p. 600; Fremy 1934, p. 143, fig. 36; Cocke 1967, p. 155, fig. 300.

Calothrix scopulorum (Weber and Mohr) C. Agardh. Plants forming a velvety, expanded, olive-green turf to a height of about 1 mm on rocks and woodwork; trichomes 8–15 μm in diameter, tapering to a hair tip, cells isodiametric or shorter than the diameter, sheaths thick, firm, yellowish-brown; heterocysts one to three in a series, basal.

C. Agardh 1824, p. 70; Bornet and Flahault 1886, p. 353; Tilden 1910, p. 258, pl. 16, fig. 11–12; Newton 1931, p. 33; Fremy 1934, p. 143, fig. 2; Lindstedt 1943, p. 40, pl. 147, fig. 39; Desikachary 1959, p. 524, pl. 111, fig. 9; Chapman 1961, p. 42, fig. 43; Cocke 1967, p. 155, fig. 301.

Calothrix aeruginea (Kützing) Thuret. Plants epiphytic or on wood or rocks, forming a light bluegreen layer about 0.5 mm tall, trichomes 7–9 μm in diameter ending in a long-tapering hair, cells much shorter than their diameter, sheaths thick colorless or yellowish; heterocysts one or two at the base and occasionally above.

Thuret 1875, p. 10; Bornet and Flahault 1886, p. 358; Tilden 1910, p. 261, pl. 17, fig. 1; Fremy 1934, p. 140, fig. 34; Lindstedt 1943, p. 38, pl. 3, fig. 9; Chapman 1961, p. 40, fig. 41; Cocke 1967, p. 154, fig. 299.

Calothrix crustacea Thuret. Plants epiphytic or on wood, stones, or shells, forming an expanded blackish-green or brownish turf 1–2 mm tall; trichomes 8–15 μm in diameter, terminating in a long hair, cells much shorter than their diameter; sheaths thick, colorless or yellowish-brown; false branches scattered, common; heterocysts one to three at the base and intercalary.

Thuret in Bornet and Thuret 1876–1880 (1876), p. 13. pl. 4; Bornet and Flahault 1886, p. 359; Tilden 1910, p. 264, pl. 17, fig. 2–6; Newton 1931, p. 34, fig. 23; Desikachary 1959, p. 523, pl. 111, fig. 10–11; Cocke 1967, p. 156, fig. 304.

Calothrix prolifera Bornet and Flahault. Forming a velvety, brownish-green plant mass; filaments 15–18 μm in diameter; 2 mm in length; some branching present; usually in the region of the heterocyst; sheaths thick, layered, colorless to yellow; trichomes 8–12 μm in diameter, tapering to a fine hair, cells wider than long; heterocysts basal and intercalary.

Growing attached to submerged wood and other objects. Cosmopolitan.

Bornet and Flahault 1886, p. 361; Collins 1906, p. 105; Tilden 1910, p. 262; Geitler 1932, p. 602; Fremy 1934, p. 147, pl. 38, fig. 3; Chapman 1956, p. 368; Cocke 1967, p. 156, fig. 303.

Calothrix vivipara Harvey.   Plants epiphytic or on stones to a height of 3–5 mm; trichomes 9–15 μm in diameter, tapering from base to apex, heterocysts basal and intercalary; sheaths thick and yellowish-brown, producing false branches in pairs.

Harvey 1858, p. 106; Bornet and Flahault 1886, p. 362; Tilden 1910, p. 263; Newton 1931, p. 35; Geitler 1932, p. 602.

Calothrix fasciculata C. Agardh.   Plants forming tufts or a turflike growth on stones and shells to a height of 2–3 mm; trichomes 8–12 microns in diameter, cells about one-half as long as the diameter or shorter, heterocysts basal and intercalary; false branches numerous; in older plants fascicles of false branches occur attached to the principal erect filaments near or above the middle.

C. Agardh 1824, p. 71; Bornet and Flahault 1886, p. 361; Collins 1891, p. 336; Tilden 1910, p. 262; Newton 1931, p. 35; Chapman 1956, p. 366, fig. 8, no. 3.

Calothrix pilosa Harvey.   Trichomes 10–20 μm in diameter, "briefly tapering at the apex, terminating in a hemispherical cell" (Tilden 1910), heterocysts basal and intercalary, the filaments generally decumbent and interwoven at the base, the upper ends erect and laterally appressed into fascicles.

Despite its lack of long-tapering tips that end in a hairlike row of cells, this species was placed in the genus *Calothrix* by Harvey and this has been accepted by later phycologists. It has been reported from Florida, the West Indies, and California where it forms a thick, dark-green pilose coating on rocks high in the intertidal zone, the plant mass to 1 cm in thickness, apparently restricted to areas of consistently high salinity. As Drouet has pointed out, *C. pilosa* is a synonym of *Scytonema hofmannii*.

Harvey 1858, p. 106, pl. 48, fig. C; Bornet and Flahault 1886, p. 363; Collins 1901, p. 242; Tilden 1910, p. 263; Collins and Hervey 1917, p. 27; Howe 1920, p. 626; Taylor 1928, p. 51; Drouet 1936, p. 21 pl. 3, fig. 17; Fremy 1938, p. 36, fig. 4; Fan 1956, p. 170, fig. 2; Joly 1957, p. 174, pl. 12, fig. 13.

Richelia

Trichomes not tapering, not terminating in a colorless hair, without a visible sheath, with a basal heterocyst; solitary or scattered inside the silica cell wall of marine plankton diatoms (*Rhizosolenia, Hemiaulus*) if the diatom wall has been cracked.

Richelia intracellularis J. Schmidt. Trichomes 5.6–9.8 μm in diameter, 50–105 μm long, heterocysts 9.8–11.2 μm in diameter, more or less spherical; cells spherical to barrel-shaped; the terminal cell often a little larger than the others.

This is a form of *Calothrix crustacea*, of worldwide distribution in temperate and tropical seas.

Schmidt 1901, p. 147; Tilden 1910, p. 201, pl. 10, fig. 8; Geitler 1932, p. 804, fig. 513; Fremy 1934, p. 188, pl. 62, fig. 3; Desikachary 1959, p. 353, pl. 61, fig. 9–11.

Tilden (1910, p. 173) lists *Calothrix rhizosolenia* Lemmermann under "Species not well understood." It has trichomes to 2.5 μm in diameter that taper slightly toward the apex within a thin sheath. It was found also in the planktonic diatoms *Rhizosolenia* and *Hemiaulus* around the Hawaiian Islands. Drouet (1973) considers *Calothrix rhizosolenia* as a form of *Calothrix crustacea*.

Fremyella

Trichomes with a basal heterocyst and sometimes with intercalary heterocycsts, tapering somewhat at the tips but not ending in a hair; the terminal cell hemispherical; one trichome in a sheath, false branching rare.

Microchaete is a synonym. All species of *Fremyella* are regarded by Drouet (1968) as forms of *Calothrix crustacea* from which the hair tips have been shed or grazed off.

The three "species" of *Fremyella* in the following key that have been recorded from marine environments are so much alike that it is impossible to separate them in a key or by means of descriptions. A fourth species that is almost identical is *Microchaete vitiensis* Askenasy reported by Fremy (1936, p. 38) from the Canary Islands and known also from Tahiti.

KEY TO THE SPECIES OF FREMYELLA

1	Trichomes 5.5–6.5 μm in diameter	*F. longifila*
1	Trichomes 5–7 μm in diameter	2
2	Trichomes 5–6 μm in diameter	*F. grisea*
2	Trichomes 6–7 μm in diameter	*F. aeruginea*

Fremyella longifila (Taylor) Drouet.   Plants   epiphytic; trichomes 5.5–6.5 μm in diameter, slightly constricted at the nodes; cells isodiametric or a little shorter than the diameter. Basal heterocysts spherical to ovate, 9–10 μm in diameter; intercalary heterocysts ovate to cylindrical, 5.6–6.5 μm in diameter, 10–12 μm long.

Taylor 1928, p. 51, pl. 2, fig. 8 (as *Calothrix longifila*); Drouet 1942, p. 130.

Fremyella grisea (Thuret) J. de Toni.   Plants forming a dense turf on shells, seagrass leaves, and algae, dull green; trichomes 5–6 μm in diameter, swollen at the base, tapering above but not hair-tipped; not constricted at the nodes, the cells shorter than the diameter; sheaths thin and colorless; heterocysts basal only, hemispherical.

Thuret in Bornet and Thuret 1880, p. 127; Tilden 1910, p. 204, pl. 10, fig. 12; Newton 1931, p. 41, fig. 29; Lindstedt 1943, p. 48, pl. 5, fig. 5 (all as *Microchaete* grisea); J. de Toni 1936, p. 4.

Fremyella aeruginea (Batters) De Toni.   Plants in tufts or forming a turf on dead corals or other solid surfaces; trichomes 6–7 μm in diameter and about 300 μm tall; not constricted at the nodes, the cells at the base about half as long as wide, somewhat longer in the upper parts of the trichomes; sheaths colorless and uniformly thick; heterocysts basal and spherical to oblong.

154

Newton 1931, p. 42; Geitler 1932, p. 666; Fremy 1934, p. 165; Desikachary 1959, p. 510, pl. 61, fig. 1–3 (all as *Microchaete aeruginea*). De Toni 1936, p. 3.

Dichothrix

Sheaths branched, false branches two to six remaining within the original sheath; heterocysts basal and intercalary. The following species are *Calothrix crustacea, sensu* Drouet (1968).

KEY TO THE SPECIES OF DICHOTHRIX

1   Plants epiphytic on larger algae   2

1   Plants not on other algae   3

2   Trichomes 15 μm in diameter, plants to 2 mm tall
                                                              D. penicillata

2   Trichomes 17–22 μm in diameter, plants 5–8 mm tall
                                                                D. fucicola

3   Trichomes 7–9 μm in diameter, plants to 1 mm tall
                                                          D. rupicola (p. 157)

3   Trichomes 6–13 μm in diameter, plants to 10 mm tall
                                                      D. bornetiana (p. 157)

Dichothrix penicillata Zenardini.   Plants usually epiphytic in tufts or
cushions; the filaments fastigiate and penicillate, about 2 mm tall, 25–35
μm in diameter; trichomes about 15 μm in diameter, cells mostly shorter
than their width. Heterocysts oblong, solitary.

This species is evidently subtropical and tropical but it drifts into colder
waters, especially on pelagic species of *Sargassum* on which it produces
the "tar spot" condition.

Zanardini 1858, p. 297, pl. 14, fig. 3; Bornet and Flahault 1886, p.
379; Collins 1901, p. 242; Tilden 1910, p. 280; Hoyt 1917–1918, p. 416;
Howe 1920, p. 626; Taylor 1928, p. 53; Rayss 1959, p. 6; Chapman 1961,
p. 39, fig. 39.

Dichothrix fucicola Bornet and Flahault.   Plants forming tufts or
small patches to a height of 5–8 mm on larger algae and stones; trichomes
9–13 μm in diameter above the somewhat swollen base where the
diameter may be 17–22 μm; the upper cells isodiametric or shorter than
the diameter; the lower cells two to three times longer than their
diameter.

Common in tropical and subtropical seas, often drifting into cooler
waters as an epiphyte on pelagic species of *Sargassum* on which it produces "tar spot."

Bornet and Flahault 1886, p. 379; Tilden 1910, p. 279; Taylor 1928, p. 52, pl. 2, fig. 15; Fremy 1938, p. 38, fig. 5; Blomquist and Humm 1946, p. 4, pl. 1, fig. 9; Chapman 1961, p. 40, fig. 40; Cocke 1967, p. 164, fig. 317.

Dichothrix rupicola Collins.   Plants in tufts about 1 mm tall, the filaments erect and penicillate; trichomes 7–9 μm in diameter, terminating in a hair; the cells mostly isodiametric.
   Collins 1901, p. 290; Tilden 1910, p. 279.

Dichothrix bornetiana Howe.   Plants producing dense, spongy, turf-like growth to a height of about 1 cm; olive-green to brownish in color, the filaments tend to be in fasciculate bundles; trichomes 6–13 μm in diameter, terminating in a hair, the cells irregular in size and shape; heterocysts basal and intercalary, elongate-ellipsoidal, 10–15 μm in diameter, 18–30 μm long.
   On stones, mud, mangrove roots, wood.
   Howe 1924, p. 357; Chapman 1961, p. 39.

Rivularia

Trichomes with a basal heterocyst, tapering at the apex to a hair tip; embedded in a common gelatinous matrix to form colonies that are hemispherical, spherical, or irregular in shape, solid, becoming hollow with age in some species. Drouet recognizes all the following as *Calothrix crustacea.*

KEY TO THE SPECIES OF RIVULARIA

1   Colonies becoming hollow with age   **2**

1   Colonies solid, not becoming hollow   **3**

2   Trichomes 2–5 μm in diameter, not constricted at nodes

                                                  R. nitida

2   Trichomes 4–5 μm in diameter below, 8–13 μm above, slightly constricted at the nodes   *R. polyotis*

3   Trichomes 3–5 μm in diameter   *R. atra* (p. 159)

3   Trichomes mostly of greater diameter   **4**

4   Trichomes 5–9 μm in diameter, cells shorter than wide

                                 R. coadunata (p. 159)

4   Trichomes mostly 4 μm in diameter, up to 16 μm; cells mostly longer than wide   *R. bornetiana* (p. 159)

Rivularia nitida C. Agardh.   Plants at first spherical or hemispherical becoming expanded, folded, corrugated, and hollow, to 3 cm wide. Trichomes 2–5 μm in diameter, terminating in a long hair; the lower cells 3–4 times as long as wide, the upper cells isodiametric or shorter than the diameter.

This species, although primarily in freshwater, has been reported from marine habitats on both coasts of North America.

Bornet and Flahault 1887, p. 359; Tilden 1910, p. 287; Howe, 1920, p. 626; Taylor 1928, p. 54, pl. 2, fig. 5; Newton 1931, p. 38; Fremy 1934, p. 154, pl. 43, fig. 2; Lindstedt 1943, p. 45, pl. 5, fig. 1; Drouet 1973, fig. 77, drawing of two trichomes from the type specimen (by designation).

Rivularia polyotis Bornet and Flahault.   Plants at first spherical or hemispherical, pulvinate, and gregarious, becoming expanded, wrinkled, and hollow to 3 cm in diameter. Trichomes 4–5 μm in diameter in the lower parts, 8.0–13.5 μm in diameter in the upper. Lower cells 1–2 times as long as wide; upper cells about one-half as long as wide, slightly to

158

strongly constricted at the cross walls. Basal heterocysts to 13 μm in diameter.

Recorded along the entire Atlantic coast of North America and the West Indies. It is epiphytic on salt marsh grasses and the aerial roots of mangroves and also on larger algae in high-salinity areas.

Bornet and Flahault 1887, p. 360; Tilden 1910, p. 286, pl. 20, fig. 5–6; Taylor 1928, p. 54, pl. 2, fig. 9.

Rivularia atra Roth.   Trichomes 2.5–5.0 μm in diameter, tapering to a hair tip and producing a basal heterocyst; embedded in a firm gelatinous matrix that is usually spherical to 4 mm in diameter, and attached to stones, wood, salt marsh plants, or larger algae.

Known from both Atlantic and Pacific coasts of North America from the Arctic to the tropics, from along the coast of England on a variety of substrata, especially in the intertidal zone. The colonies may become confluent and form a layer. Cosmopolitan.

Bornet and Flahault 1887, p. 353; Tilden 1910, p. 289, pl. 20, fig. 10; Newton 1931, p. 38, fig. 26; Fremy 1934, p. 153, pl. 42, fig. 2; 1936, p. 35; Lindstedt 1943, p. 45, pl. 5, fig. 2, 7; Chapman 1961, p. 45, fig. 49; Drouet 1973, fig. 75, drawing of two trichomes from the type specimen (designated).

Rivularia coadunata (Sommerfelt) Foslie.   Colonies hemispherical when young, becoming irregular and cushionlike with age 2–8 mm in thickness; trichomes 5–9 μm in diameter, terminating in a long, slender hair; cells shorter than their diameter; heterocysts one to three at the base, sometimes intercalary also hemispherical to oblong.

This species is found on rocks in fresh-, brackish, and saltwater.

Foslie 1891, p. 21; Tilden 1910, p. 291, pl. 20, fig. 16–17.

Rivularia bornetiana Setchell.   Colonies firm, solid, spherical to cylindrical, olive-green or yellow-green, often encrusted with calcium carbonate; trichomes 7–16 μm in diameter, radiating from the center and having distinct individual sheaths; cells one-half to five times longer than the diameter; basal heterocysts spherical-depressed or ellipsoidal, 6–8 μm in diameter.

Setchell 1895, p. 426; Collins 1900, p. 43; Tilden 1910, p. 292.

Scytonema

The description of the genus *Scytonema* that follows is that of the older literature, prior to Drouet (1973).

Trichomes single within the sheath, false branches abundant, single or in pairs; apex of trichome slightly tapered but not hair-tipped, the apical cell hemispherical, sometimes slightly swollen.

Apparently only one species of *Scytonema* of the older literature has been recorded from a truly marine habitat. Another species that grows in a marine habitat was known in the older literature as *Calothrix pilosa*. It is puzzling that it was placed in the genus *Calothrix* since it does not have the one basic characteristic of that genus, the tapering of the apex into a hair.

Scytonema ocellatum Lyngbye was reported from a slightly brackish pool at Dry Tortugas, Florida, by Taylor (1928, p. 49). *Scytonema tolypothrichoides* Kützing was reported from a brackish-water stream by Chapman (1956, p. 365). These species are normally found on damp soil, wood, or stones.

Scytonema fuliginosum Tilden.   Trichomes 10–20 μm in diameter; cells 1.4–5.0 μm long; heterocysts 12–16 μm in diameter, spherical to ovate; sheaths thick, becoming brown with age.

This species was described from a tide pool on the island of Hawaii. It is a synonym of *S. hofmannii* (Drouet 1973).

Tilden 1910, p. 225, pl. 13, fig. 7–8.

Hormothamnium

Trichomes with intercalary heterocysts, not attached at the base, not tapering, forming an expanded layer of confluent sheaths in which the individual sheaths are indistinct, some sheaths appearing to have several trichomes.

The genera *Hormothamnium*, *Nodularia*, *Anabaena*, and *Nostoc* cannot be separated adequately even on the basis of the environmentally variable characters used in the older literature to establish them.

Two species of *Hormothamnium* have been repeatedly recorded from marine habitats. Both are synonyms of *Anabaina oscillarioides* (Drouet 1978).

KEY TO THE SPECIES OF HORMOTHAMNIUM

Trichomes 6–7 μm in diameter   *H. enteromorphoides*

Trichomes 9–12 μm in diameter   *H. solutum*

Hormothamnium enteromorphoides Grunow.   Trichomes 6–7 μm in diameter; cells a little longer than wide, the heterocysts slightly larger and ovate; sheaths soft and adherent, with one to several trichomes, if distinct; colonies forming a flat layer of confluent sheaths or a branched structure under some conditions in remote resemblance of a small *Enteromorpha* plant.

Widely distributed in tropical seas.

Grunow 1867, p. 31, pl. 1, fig. 2; Tilden 1910, p. 205, pl. 10, fig. 13; Taylor 1928, p. 48, pl. 2, fig. 3; Fremy 1938, p. 44, fig. 7; Desikachary 1959, p. 433, pl. 61, fig. 6–8; Chapman 1961, p. 38, fig. 38.

Hormothamnium solutum Bornet and Grunow.   Trichomes 9–12 μm in diameter; cells disk-shaped, three to four times as wide as long; heterocysts isodiametric. Extracellular polysaccharides forming a firm sheath, these adhering laterally to form fascicles that are usually erect and 5–6 mm high.

Known from Hawaii (the type locality), India, and the Caribbean Sea.

Bornet and Flahault 1888, p. 259; Tilden 1910, p. 205; Desikachary 1959, p. 433.

Anabaina

Trichomes 2.5–14 μm in diameter; the cells spherical to barrel-shaped, isodiametric, or slightly shorter or longer than the diameter; heterocysts numerous, intercalary and basal; spores usually larger and more elongate than the vegetative cells, single or in a series between heterocysts. Extracellular polysaccharide thin and invisible, soft and diffluent, or trichomes without an apparent sheath.

The genus is primarily a freshwater one, but a few species are reported from seawater. It is similar to *Nostoc* but has diffluent extracellular polysaccharides that do not accumulate as a gelatinous matrix.

161

The spelling "*Anabaena*" is an error. The genus was originally described as *Anabaina* by Bory de Saint-Vincent in 1822. A few years later, another author used the spelling "*Anabaena*" and that spelling has been used in virtually all publications since then. "*Anabaena*" is not available for conservation.

Drouet places the following three species in *Anabaina oscillarioides*.

KEY TO THE SPECIES OF ANABAINA

1 Heterocysts spherical   *A. inaequalis*

1 Heterocysts somewhat elongate to cylindrical   **2**

2 Heterocysts becoming ovate, spores not adjacent to heterocysts
                                        A. variabilis

2 Heterocysts becoming cylindrical, spores adjacent to heterocysts
                                        A. torulosa

Anabaina inaequalis (Kützing) Trevisan. Trichomes 4–5 μm in diameter; the sheath invisible or visible and distinct, or the sheaths fused to form a stratum; cells spherical to subspherical; heterocysts about 6 μm in diameter, spherical; spores 6–8 μm in diameter, 14–17 μm long, remote from heterocysts; 2–3 in a series, the wall yellowish with maturity.
     Of worldwide distribution in fresh-and saltwater.
     Bornet and Flahault 1888, p. 231; Tilden 1910, p. 191, pl. 9, fig. 16; Cocke 1967, p. 100, fig. 218.

Anabaena variabilis Kützing. Trichomes 4–6 μm in diameter, spherical or a little longer than the diameter, the apical cell obtuse to conical; heterocysts 6 μm in diameter, 8–14 μm long, blunt at the ends, spores in series remote from the heterocysts, the wall smooth and yellow-brown when mature. Extracellular polysaccharide thin, invisible.
     Known along both the Atlantic and Pacific coasts of North America, from the British Isles and New Zealand, and in brackish and seawater (but also in freshwater).
     Kützing 1843, p. 210; Bornet and Flahault 1888, p. 226; Tilden 1910, p. 187, pl. 9, fig. 9; Newton 1931, p. 44; Geitler 1932, p. 886, fig. 567e; Fremy 1934, p. 182, pl. 61; Linstedt 1943, p. 52, pl. 6, fig. 6–7, 15; Chapman 1956, p. 362, fig. 9, no. 3; Cocke 1967, p. 97, fig. 207.

Anabaina torulosa (Carmichael) Lagerheim. Trichomes 4–5 μm in diameter, the cells tending to be spherical but with flattened ends from mutual pressure, the apical cell acute-conical; heterocysts 6 μm in diameter, 6–10 μm long, spherical to oval; spores relatively large, 7–17

μm in diameter, 18–28 μm long, oval to cylindrical, single or in series contiguous to heterocysts. Extracellular polysaccharides diffluent and hold the trichomes in layers or masses.

Apparently of worldwide distribution in fresh- and brackish waters, in salt marshes, and sometimes in full seawater.

Lagerheim 1883, p. 47; Bornet and Flahault 1888, p. 236; Tilden 1910, p. 192, pl. 9, fig. 19; Newton 1931, p. 45, fig. 31; Fremy 1936, p. 40; Lindstedt 1943, p. 52, pl. 6, fig. 3–5; Desikachary 1959, p. 415, pl. 71, fig. 6; Cocke 1967, p. 99, fig. 213.

Nostoc

The following description is of the genus *Nostoc* as interpreted prior to publication of Drouet's *Revision of the Nostocaceae with Constricted Trichomes* in 1978.

Trichomes embedded in a globose, oblong, or irregularly shaped gelatinous matrix that may be solid or hollow, attached or loose, and in which the trichomes are curved, flexuous, entangled; cells depressed-spherical, spherical, moniliform, or barrel-shaped; heterocysts intercalary, terminal in young plants; spores spherical to oblong, produced centrifugally in series between the heterocysts.

Nostoc is a freshwater genus. Newton (1931) recorded two species from marine habitats. Chapman (1956) recorded these two and three additional from New Zealand marine sources. The two species recorded by both these authors are treated in the key and descriptions that follow. The three additional species recorded by Chapman, *N. commune* Vaucher, *N. sphaericum* Vaucher, and *N. microscopicum* Carmichael, are omitted.

Trichomes 2.5–3.0 μm in diameter; plants microscopic, within the polysaccharide of other algae   *N. entophytum*

Trichomes 3.5–4.0 μm in diameter; macroscopic, not endophytic
                                                              N. linckia

Nostoc entophytum Bornet and Flahault.   Plant minute, invisible without magnification, densely aggregate; trichomes twisted, 2.5–3.0 μm in diameter, the cells spherical or flattened; heterocysts somewhat larger, spores 5–6 μm in diameter, usually rounded.

Found growing within the polysaccharide of other algae. Reported from the coast of Britain and from New Zealand.

Drouet (1978) lists this name as a synonym of *Nostoc commune*, a freshwater species that is rarely found in brackish water or in habitats occasionally inundated by seawater.

Bornet and Flahault 1886–1888 (1888), p. 190; Newton 1931, p. 44; Chapman 1956, p. 361; Desikachary 1959, p. 375.

Nostoc linckia Bornet.   Colonies spherical gelatinous, filaments entangled; trichomes 3.4–4.0 μm in diameter, cells spherical to flattened on adjacent faces; heterocysts 5–6 μm in diameter; spores in series, usually on each side of a heterocyst.

This species is more commonly reported from freshwater, but has been found in marine habitats in the British Isles and New Zealand. Drouet (1978) lists this name under *nomina excludenda* as the type material is *Calothrix parietina*.

Bornet in Bornet and Flahault 1886–88 (1888), p. 192; Newton 1931, p. 44, fig. 30; Geitler 1932, p. 838, fig. 528b; Chapman 1956, p. 362, fig. 8, no. 6.

Nodularia

Filaments free, extracellular polysaccharides not in evidence or embedded in a mucilaginous matrix; much constricted at the cross walls, often moniliform; cells isodiametric or shorter than the diameter, discoid to spherical-depressed; heterocysts intercalary in origin, terminal by fragmentation of the trichome. Spores develop serially between the heterocysts, but may be absent under some conditions.

KEY TO THE SPECIES OF NODULARIA

1 Trichomes 8–18 μm in diameter, cells shorter than wide; spores spherical   *N. spumigena*

1 Cells mostly isodiametric, spores elliptical   **2**

2 Trichomes 4–6 μm in diameter   *N. harveyana*

2 Trichomes 7.5–9.0 μm in diameter   *N. hawaiiensis*

Nodularia spumigena Mertens.   Trichomes 8–18 μm in diameter, cells discoid, three or four times shorter than their diameter; heterocysts discoid, of slightly greater dimensions than the vegetative cells; spores 12–15 μm in diameter, 6–12 μm long, not contiguous to the heterocysts; sheaths delicate or fused into a stratum.

Widely distributed in the sea in salt marshes and in quiet, shallow water on bottom sediments. This name is a synonym of *Nostoc spumigena* (Drouet 1978).

Mertens in Jürgens 1816–1822 (1822), folio 15, no. 4; Tilden 1910, p. 184, pl. 9, fig. 7–8; Newton 1931, p. 46, fig. 32A–C; Lindstedt 1943, p. 50, pl. 6, fig. 1–2; Cocke 1967, p. 88, fig. 191.

Nodularia harveyana (Thwaites) Thuret.   Trichomes 4–6 μm in diameter in a thin, colorless sheath or the sheaths confluent; cells isodiametric or shorter than the diameter, heterocysts about the same size; spores spherical, compressed, 6–8 μm in diameter, the wall colorless or light brown.

Widely distributed on sand or muddy sand in the intertidal zone in salt marshes, estuaries, bays, or protected places. *N. harveyana* is considered a synonym of *Nostoc spumigena* by Drouet (1978).

Thuret 1875, p. 378; Tilden 1910, p. 182, pl. 9, fig. 1–2; Lindstedt 1943, p. 50, pl. 5, fig. 6; pl. 6, fig. 8–9; Humm and Caylor 1957, p. 240, pl. 2, fig. 1; Cocke 1967, p. 88, fig. 190.

Nodularia hawaiiensis Tilden.   Trichomes 7.5–9.5 μm in diameter, cells mostly spherical before division; heterocysts 10 μm in diameter, spherical or a little longer than the diameter. Extracellular polysaccharides invisible.

Oshu, Hawaii, in tufts on larger algae on a wave-beaten reef. This species is regarded as a synonym of *Anabaina oscillarioides* by Drouet (1978).

Tilden 1910, p. 184, pl. 9, fig. 5.

Table 2   List of Species from the Older Literature Treated in this Work and the Valid name of Each, sensu Drouet

Older Name	Valid Name, *sensu* Drouet
Amphithrix janthina	*Schizothrix calcicola*
Amphithrix violacea	*Schizothrix calcicola*
Anabaena inaequalis	*Anabaina oscillarioides*
Anabaena torulosa	*Anabaina oscillarioides*
Anabaena variabilis	*Anabaina oscillarioides*
Aphanocapsa marina	*Anacystis marina*
Aphanothece pallida	*Coccochloris stagnina*
Calothrix aeruginea	*Calothrix crustacea*
Calothrix confervicola	*Calothrix crustacea*
Calothrix consocieta	*Calothrix crustacea*
Calothrix contarenii	*Calothrix crustacea*
Calothrix crustacea	*Calothrix crustacea*
Calothrix fasciculata	*Calothrix crustacea*
Calothrix fusco-violacea	*Calothrix crustacea*
Calothrix parasitica	*Calothrix crustacea*
Calothrix pilosa	*Scytonema hofmannii*
Calothrix prolifera	*Calothrix crustacea*
Calothrix pulvinata	*Calothrix crustacea*
Calothrix scopulorum	*Calothrix crustacea*
Calothrix vivipara	*Calothrix crustacea*
Chroococcus minutus	*Coccochloris stagnina*
Chroococcus turgidus	*Anacystis dimidiata*
Dermocarpa biscayensis	*Entophysalis conferta*
Dermocarpa leibleiniae	*Entophysalis conferta*
Dermocarpa minima	*Entophysalis conferta*
Dermocarpa olivacea	*Entophysalis conferta*
Dermocarpa prasina	*Entophysalis conferta*
Dermocarpa rosea	*Entophysalis conferta*
Dermocarpa smaragdina	*Entophysalis conferta*
Dermocarpa solitaria	*Entophysalis conferta*

Table 2. *Continued*

Older Name	Valid Name, *sensu* Drouet
Dermocarpa violacea	*Entophysalis deusta*
Dichothrix bornetiana	*Calothrix crustacea*
Dichothrix fucicola	*Calothrix crustacea*
Dichothrix penicillata	*Calothrix crustacea*
Dichothrix rupicola	*Calothrix crustacea*
Entophysalis granulosa	*Entophysalis deusta*
Entophysalis magnoliae	*Anacystis montana forma montana*
Entophysalis violacea	*Entophysalis deusta*
Fremyella aeruginea	*Calothrix crustacea*
Fremyella grisea	*Calothrix crustacea*
Fremyella longifila	*Calothrix crustacea*
Gloeocapsa crepidinum	*Entophysalis deusta*
Hormothamnium enteromorphoides	*Anabaina oscillarioides*
Hormothamnium solutum	*Anabaina oscillarioides*
Hydrocoleum cantharidosmum	*Microcoleus lyngbyaceus*
Hydrocoleum comoides	*Microcoleus lyngbyaceus*
Hydrocoleum glutinosum	*Microcoleus lyngbyaceus*
Hydrocoleum lyngbyaceum	*Microcoleus lyngbyaceus*
Hyella caespitosa	*Entophysalis deusta*
Lyngbya aestuarii	*Microcoleus lyngbyaceus*
Lyngbya confervoides	*Microcoleus lyngbyaceus*
Lyngbya epiphytica	*Schizothrix calcicola*
Lyngbya gracilis	*Schizothrix mexicana*
Lyngbya majuscula	*Microcoleus lyngbyaceus*
Lyngbya meneghiniana	*Schizothrix mexicana*
Lyngbya rivulariarum	*Schizothrix calcicola*
Lyngbya semiplena	*Microcoleus lyngbyaceus*
Lyngbya sordida	*Schizothrix mexicana*
Merismopedia elegans variety marina	*Agmenellum thermale*
Merismopedia glauca forma mediterranea	*Agmenellum thermale*
Merismopedia punctata forma minor	*Agmenellum quadruplicatum*
Merismopedia sabulicola	*Agmenellum thermale*
Merismopedia warmingiana	*Agmenellum quadruplicatum*
Microcoleus chthonoplastes	*Schizothrix arenaria*
Microcoleus tenerrimus	*Schizothrix tenerrima*
Nodularia harveyana	*Nostoc spumigena*
Nodularia hawaiiensis	*Anabaina oscillarioides*
Nodularia spumigena	*Nostoc spumigena*
Nostoc entophytum	*Nostoc commune*
Nostoc linckia	*Nominum excludendum*

169

Table 2. *Continued*

Older Name	Valid Name, *sensu* Drouet
Oncobyrsa marina	Entophysalis conferta
Oscillatoria acuminata	Schizothrix tenerrima
Oscillatoria amoena	Microcoleus vaginatus
Oscillatoria bonnemaisonii	Microcoleus lyngbyaceus
Oscillatoria brevis	Arthrospira neapolitana
Oscillatoria corallinae	Microcoleus lyngbyaceus
Oscillatoria laetevirens	Schizothrix arenaria
Oscillatoria margaritifera	Microcoleus lyngbyaceus
Oscillatoria miniata	Microcoleus lyngbyaceus
Oscillatoria nigro-viridis	Porphyrosiphon notarisii
Oscillatoria salinarum	Schizothrix arenaria
Oscillatoria subuliformis	Porphyrosiphon notarisii
Phormidium autumnale	Microcoleus vaginatus
Phormidium crosbyanum	Schizothrix calcicola
Phormidium fragile	Schizothrix calcicola
Phormidium penicillatum	Oscillatoria submembranacea
Phormidium persicinum	Schizothrix calcicola
Phormidium spongeliae	Schizothrix mexicana
Phormidium submembranaceum	Oscillatoria submembranacea
Phormidium subuliforme	Schizothrix arenaria
Plectonema battersii	Schizothrix calcicola
Plectonema calothrichoides	Schizothrix calcicola
Plectonema nostocorum	Schizothrix calcicola
Plectonema terebrans	Schizothrix calcicola
Pleurocapsa amethystea	Entophysalis conferta
Pleurocapsa crepidinum	Entophysalis deusta
Pleurocapsa fuliginosa	Entophysalis deusta
Richelia intracellularis	Calothrix crustacea
Rivularia atra	Calothrix crustacea
Rivularia bornetiana	Calothrix crustacea
Rivularia coadunata	Calothrix crustacea
Rivularia nitida	Calothrix crustacea
Rivularia polyotis	Calothrix crustacea
Schizothrix lacustris	Schizothrix calcicola
Schizothrix longiarticulata	Schizothrix arenaria
Scytonema fuliginosum	Scytonema hormannii
Spirulina major	Spirulina subsalsa
Spirulina meneghiniana	Spirulina subsalsa
Spirulina nordstedtii	Spirulina subsalsa
Spirulina subsalsa	Spirulina subsalsa

Table 2. *Continued*

Older Name	Valid Name, *sensu* Drouet
Spirulina subtilissima	*Spirulina subsalsa*
Spirulina tenerrima	*Spirulina subsalsa*
Spirulina versicolor	*Spirulina subsalsa*
Symploca atlantica	*Oscillatoria submembranacea*
Symploca hydnoides	*Nominum excludendum*
Symploca laete-viridis	*Schizothrix arenaria*
Trichodesmium contortum	*Oscillatoria erythraea*
Trichodesmium erythraeum	*Oscillatoria erythraea*
Trichodesmium thiebautii	*Oscillatoria erythraea*
Xenococcus acervatus	*Entophysalis conferta*
Xenococcus schousboei	*Entophysalis conferta*

GLOSSARY

acuminate	tapering gradually to a point
agglutinate	mutually adhering by extracellular polysaccharide
apical cell	terminal cell of a trichome
attenuate	tapering gradually
benthic	in an aquatic habitat, attached to a substrate
caespitose	growing in clumps or turflike masses, matted
capitate	swollen at the end
cartilaginous	firm and tough but flexible
catenate	connected into a chainlike series
coalesced	grown together, united
coccoid	more or less spherical
confluent	fusing, running together
constricted	abruptly contracted
contiguous	in contact
decumbent	prostrate but the tips bent upward
diffluent	dissolving away, becoming fluid
discoid	disklike
endospore	small cells resulting from internal division and retained for a time within the parent cell wall
endosporangium	wall and contents of a cell that has divided internally into endospores
endolithic	penetrating rock (here, principally limestone)
endophytic	penetrating a plant, but not necessarily parasitic
epiphytic	growing upon or attached to a plant
false branch	formed when a trichome grows through the side wall of the sheath and then forms another sheath around itself; a condition in which the sheath is branched but not the trichomes
fascicle	a cluster of parallel filaments

fasciculate	in bundles
fastigiate	fasciculate filaments or trichomes that taper to a point as a group
filament	a trichome and its sheath
globose	more or less spherical
gonidium (gonidia)	an old term for endospores or small cells of coccoid bluegreen algae
hormogonium (hormogone)	a short segment of a trichome
hyaline	transparent
intercalary	occurring between the ends, between cells
isodiametric	of approximately the same diameter in all dimensions
lamellose	having layers
littoral	intertidal; in shallow water near shore
lyse	to rupture or burst
matrix	polysaccharide sheath or envelope surrounding a group of cells or trichomes
membranaceous	thin, papery
moniliform	like a string of beads; deeply constricted at the nodes
node	juncture of two cells of a trichome
obtuse	rounded or blunt
ovoid	egg-shaped
pelagic	occurring unattached in the open sea
pellucid	clear, transparent
pilose	hairy
plankton	organisms living in suspension in the water and transported by ocean currents
plicate	folded
polyhedral	having many sides
procaryote	an organism having a primitive form of cellular organization in which there is no organized nucleus, no mitosis, and the photosynthetic pigments are not enclosed in plastids

pseudovacuoles	gas-filled cell inclusions, usually rod-shaped and temporary, with an extremely thin membrane
pulvinate	cushion-shaped
pyriform	pear-shaped
quadrate	squarish
rectilinear	arranged in rows in two directions
rotund	rounded
septate	having cross walls; divided into cells
sheath	polysaccharide coating around a cell, trichome or colony
spore	one-celled reproductive body
stratum	a layer
substrate	surface to which a plant is attached
torulose	twisted or knobby; irregularly swollen
trichome	in filamentous bluegreen algae, a row of cells without the sheath
truncate	squared off at the end, blunt
uniseriate	arranged in a single row

BIBLIOGRAPHY

Agardh, C. A. 1817. Synopsis Algarum Scandinaviae, Adjecta Dispositione Universale Algarum. Lund. xl + 135 p.

Agardh, C. A. 1824. Systema Algarum. Lund. xxxviii + 312 p.

Ahlborn, F. 1895. Über die Wasserblüte *Byssus Flos-aquae* und ihr Verhalten gegen Druck. Verh. Naturwiss, Ver. Hamburg III 2:25.

Allen, M. M. 1968. Photosynthetic membrane system in *Anacystis nidulans.* J. Bact. 96:836–841.

Almodovar, L. R. and H. L. Blomquist. 1961. Notes on the marine algae of Cabo Rojo, Puerto Rico. Quart. J. Florida Acad. Sci. 24:80–93.

Asato, Y. and C. E. Folsome. 1970. Temporal genetic mapping of the bluegreen alga *Anacystis nidulans.* Genetics 65:407–419.

Aziz, K. M. S. and H. J. Humm. 1962. Additions to the algal flora of Beaufort, N. C., and vicinity. J. Elisha Mitchell Sci. Soc. 78:55–63.

Baker, A. F. and H. C. Bold. 1970. Phycological studies. X. Taxonomic studies in the Oscillatoriaceae. Univ. of Texas Publ. No. 7004, Austin. 105 pp., 125 figs.

Ballantine, D. and H. J. Humm. 1975. Benthic algae of the Anclote estuary. I. Epiphytes of seagrass leaves. Florida Sci. 38:149–162.

Batters, E. A. L. 1889. Marine algae of Berwick. Hist. Berwickshire Nat. Club, Vol. 12.

Bazin, M. J. 1968. Sexuality in a bluegreen alga: Genetic recombination in *Anacystis nidulans.* Nature 218:282–283.

Bennett, A. and L. Bogorad. 1971. Properties of subunits and aggregates of bluegreen algal biliproteins. Biochemistry 10:3625–3634.

Bernard, F. 1963. Density of flagellates and Myxophyceae in the heterotrophic layers related to environment. *In* C. Oppenheimer (Editor), Symposium on Marine Microbiology, p. 215–218. Charles C. Thomas, Springfield, IL.

Blackwelder, B. C. 1975. Attached forms of intertidal algae from the coastal regions of South Carolina. Ph.D. thesis, Univ. of South Carolina. 322 pp., 117 figs., 46 tables.

Blomquist, H. L. and H. J. Humm. 1946. Some marine algae new to Beaufort, N. C. J. Elisha Mitchell Sci. Soc. 62:1–8.

Bornet, E. and C. Flahault. 1886–88. Revision des Nostocacées Heterocystées contenues dans les principaux herbiers de France, I–IV. Ann. Sci. Nat., Bot., Ser. 7, 3:323–81; 4:343–73; 5:51–129; 7:177–262.

Bornet, E. and C. Flahault. 1889. Sur quelques plantes vivant dans le test calcaire des Mollusques. Bul. Soc. Bot. France 36:147–176.

Bornet, E., and G. Thuret. 1876–1880. Notes algologique; recueil d'observation sur les algues. I:1–72, pl. 1–25, 1876; II:73–196, pl. 26–50, 1880. Paris.

Bory de St. Vincent, J. B. 1822–1831. Dictionnaire Classique d'Histoire Naturelle. Vol. 1–17. Paris.

Bowen, C. C. and T. E. Jensen. 1965. Bluegreen algae: Fine structure of the gas vacuoles. Science 147:1460–1462.

Brandt, F. 1903. Nordisches Plankton. 20:2.

Breed, R. S., R. G. E. Murray, and N. R. Smith (Editors). 1957. Bergey's Manual of Determinative Bacteriology, 7th edition, Baltimore. 1132 pp.

Bristol-Roach, B. M. 1919. On the retention of vitality by algae from old stored soils. New
Phytol. 18:92–107.

Bristol-Roach, B. M. 1920. On the algal flora of some dessicated English soils. Ann. Bot. 34:
35.

Buchanan, R. E. and N. E. Gibbons (Editors). 1974. Bergey's Manual of Determinative
Bacteriology, 8th edition, Baltimore, 1246 pp.

Carpenter, E. J. and C. C. Price. 1976. Marine *Oscillatoria* (*Trichodesmium*): Explanation
for aerobic nitrogen fixation without heterocysts. Science 191:1278–80.

Chapman, V. J. 1956. The marine algae of New Zealand. J. Linn. Soc. London 55:348–373;
55:333–501.

Chapman, V. J. 1961. The marine algae of Jamaica. Part I, Myxophyceae and Chlorophy-
ceae. Institute of Jamaica, Kingston. 159 pp., 176 figs.

Cocke, E. C. 1967. The Myxophyceae of North Carolina. Publ. by the author, Wake Forest
Univ., Winston-Salem, N.C. i-vii, 206 pp., 326 figs.

Cohen, Y., B. B. Jorgensen, E. Padan, and M. Shilo. 1975. Sulfide-dependent anoxygenic
photosynthesis in the cyanobacterium *Oscillatoria limnetica*. Nature 257:489–92.

Collins, F. S. 1891. Notes on New England marine algae, V. Bull. Torrey Bot. Club 18:335–
341.

Collins, F. S. 1896. Notes on New England marine algae, VI. Bull. Torrey Bot. club 23:

Collins, F. S. 1900. Preliminary list of New England plants. V. Marine algae. Rhodora 2:43.

Collins, F. S. 1901. The algae of Jamaica. Proc. Amer. Acad. Arts Sci. 37:231–270.

Collins, F. S. 1906. New species, etc., issued in the Phycotheca Boreali-Americana. Rhodora
8:104–113.

Collins, F. S. and A. B. Hervey. 1917. The algae of Bermuda. Proc. Amer. Acad. Arts and
Sci. 53:1–195, 6 pl.

Crouan, P. L., and H. M. Crouan. 1858. Note sur quelques algues marines nouvelle de la
rode de Brest. Ann. Sci. Nat., Bot., ser. 4, 9:69–75.

Crouan, P. L. and H. M. Crouan. 1860. Liste des algues marine decouvertes dans le Finistere. Bull. Soc. Bot. de France 7:367–373, 836–839.

Crouan, P. L. and H. M. Crouan. 1867. Florule du Finistere. Paris. x + 262 pp., 32 pl.

Daft, M. J., and W. D. P. Stewart. 1971. Bacterial pathogens of bluegreen algae. New
Phytol. 70:819–829.

Dawes, C. J. 1974. Marine Algae of the West Coast of Florida. Univ. of Miami Press. 201 pp.,
82 figs.

Delaney, S. F., and N. G. Carr. 1975. Temporal genetic mapping in the bluegreen alga *Anacystis nidulans* using ethyl methane sulphonate. J. Gen. Microbiol. 88:259–68.

Desikachary, T. V. 1959. Cyanophyta. Indian Council of Agricultural Research, New Delhi.
686 pp., 139 pl.

De Toni, G. 1936. Notorelle di nomenclature algologica, VIII. Primo elenco di Floridee
omonime. Privately printed. Brescia.

De Toni, G. 1938. A note on phycological nomenclature. Rhodora 40:27.

Donze, M., J. Haveman, and P. Schiereck. 1972. Absence of photosystem 2 in heterocysts of the bluegreen alga *Anabaena*. Biochim. Biophys. Acta 256:157–161.

Drewes, K. 1928. Uber die Assimilation des Luftstickstoffs durch Blaualgen. Centrabl. Bakt. Parasit., II, 76:88–101.

Drouet, F. 1936. Myxophyceae of the G. Allan Hancock expedition of 1934, collected by William Randolph Taylor. Allan Hancock Pacific Exped. 3(2):15–31.

Drouet, F. 1942. The filamentous Myxophyceae of Jamaica. Publ. Field Mus., Bot. Ser., 20(5):107–22.

Drouet, F. 1951. Cyanophyta. *In* G. M. Smith (Editor), Manual of Phycology, Chapt. 8, pp. 159–166. Chronica Botanica, Waltham, Mass.

Drouet, F. 1962. Gomont's ecophenes of the bluegreen alga, *Microcoleus vaginatus* (Oscillatoriaceae). Proc. Acad. Nat. Sci. Philadelphia 114:191–205.

Drouet, F. 1963. Ecophenes of *Schizothrix calcicola* (Oscillatoriaceae). Proc. Acad. Nat. Sci. Philadelphia 115:261–281.

Drouet, F. 1968. Revision of the Classification of the Oscillatoriaceae. Monograph 15, Acad. Nat. Sci. Philadelphia. 334 pp., 131 figs.

Drouet, F. 1969. Homonymy in *Arthrospira* Stizenberger (Oscillatoriaceae). Phycologia 18: 339.

Drouet, F. 1973. Revision of the Nostocaceae with Cylindrical Trichomes. Hafner Press, New York, 292 pp., 83 figs.

Drouet, F. 1978. Revision of the Nostocaceae with Constricted Trichomes. J. Cramer, Vaduz, Liechtenstein. 258 pp., 3 pl.

Drouet, F. and W. A. Daily. 1952. A synopsis of the coccoid Myxophyceae. Butler Univ. Bot. Studies 10:120–23.

Drouet, F. and W. A. Daily. 1956. Revision of the coccoid Myxophyceae. Butler Univ. Bot. Studies 13:1–218.

Dugdale, R. C., J. J. Goering, and J. H. Ryther. 1964. High nitrogen fixation rates in the Sargasso Sea and the Arabian Sea. Limnol. Oceanogr. 12:196–206.

Edwards, M. R. and E. Gantt. 1971. Phycobilisomes of the thermophilic bluegreen alga *Synecococcus lividus*. J. Cell Biol. 50:896–900.

Ehrenberg, C. G. 1830. Neue Beobachtungen uber blutartige Erscheinungen in Aegypten, Arabien und Sibirien. Ann. Physik. Chem., Ser. 2, 18:477–514.

Evans, E. L. and M. M. Allen. 1973. Phycobilisomes in *Anacystic nidulans*. J. Bact. 113:403–408.

Fan, K. C. 1956. Revision of *Calothrix*. Revue Algologique, New Ser., 2(3):154–178.

Farlow, W. G. 1881. The marine algae of New England. Report, U.S. Commissioner of Fish and Fisheries for 1879, appendix A-1:1–210.

Fay, P. 1970. Photostimulation of nitrogen fixation in *Anabaena cylindrica*. Biochim. Biophys. Acta 216:353–356.

Feldmann, J. 1933. Sur quelques Cyanophycees vivant dans le tissu des eponges de Banyuls. Arch. Zool. Exp. Gen., Jub. Vol. 75:381–404.

Fogg, G. E. 1952. The production of extracellular nitrogenous substances by a bluegreen alga. Proc. Roy. Soc. (London) B139:372–397.

Fogg, G. E. 1956. The comparative physiology and biochemistry of the bluegreen algae. Bact. Rev. 20:148–165.

Forti, A. 1907. Sylloge Myxophycearum omnium hucusque cognitarum.. In J. B. De Toni (Editor), Sylloge Algarum, Vol. 5, pp. 1–761. Padova.

Foslie, M. 1890–1891. Contribution to knowledge of the marine algae of Norway, I and II. Tromsø. Mus. Aarsh., Vol. 13 and 14. Tromsø.

Frank, B. 1889. Uber den experimentellen Nachweis der Assimilation freien Stickstoffs durch erdbewohnende Algen. Ber. Deut. Bot. Ges. 7:34–42.

Frank, H., M. Lefort, and H. H. Martin. 1962. Elektronenoptische und chemische Untersuchungen an Zellwanden der Blaualge Phormidium uncinatum. Z. Naturforsch. 17b:262–268.

Fremy, P. 1929. Les Myxophycees de l'Afrique equatoriale francaise. Arch. Bot., Caen, Vol. 3, memoire 2.

Fremy, P. 1934. Cyanophycees des cotes d'Europe. Mem. Soc. Sci. Nat. Math. Cherbourg 41:1–234, 66 pl.

Fremy, P. 1936. Marine algae from the Canary Islands. IV. Cyanophyceae. Det. Kgl. Danske Vidensk. Selsk., Biol. Meddel. 12(5):1–43.

Fremy, P. 1938. Cyanophycees marines des anciennes Antilles Danoises. Dansk Bot. Arkiv 9(7):1–46.

Fries, E. 1835. Corpus florarum provincialium Sueciae. I. Floram scanica. Upsala. xxiv + 394 pp.

Fritsch, F. E. 1951. The heterocyst: A botanical enigma. Proc. Linn. Soc. London 162:194–211.

Fuhs, G. W. 1958. Bau, Verhalten und Bedeutung der kernaquivalenten Structuren bei Oscillatoria amoena. Arch. Mikrobiol. 28:270–302.

Fuhs, G. W. 1969. The nuclear structures of protocaryotic organisms (bacteria and Cyanophyceae). Protoplasmatologia 5:1–186.

Gallon, J. R., T. A. LaRue, and W. G. W. Kurz. 1974. Photosynthesis and nitrogenase activity in the blue-green alga, Gloeocapsa. Can. J. Microbiol. 20:1633–38.

Gardner, N. L. 1918a. New Pacific coast marine algae II. Univ. Calif. Publ. Bot. 6(16):429–454.

Gardner, N. L. 1918b. New Pacific coast marine algae III. Univ. Calif. Publ. Bot. 6(17):455–486.

Gardner, N. L. 1927. New Myxophyceae from Puerto Rico. Mem. N.Y. Bot. Gard. 7:1–144.

Gardner, N. L. 1932. Myxophyceae of Puerto Rico and the Virgin Islands. Sci. Survey Puerto Rico Virgin Islands 8(2):249–311. New York Acad. Sci.

Geitler, L. 1932. Cyanophyceae. In L. Rabenhorst (Editor), Kryptogamen-Flora von Deutschland, Osterreich und der Schweiz. Second edition, Vol. 14, pp. 1–1196. G. Fischer, Jena.

Geitler, L. 1942. Schizophyceae. In A. Engler and K. Prantl, Die Naturlichen Pflanzenfamilien. Second edition. 1b:1–232, 156 fig. Leipzig.

Goering, J. J., R. C. Dugdale, and D. W. Menzel. 1966. Estimates of nitrogen uptake by *Trichodesmium* sp. in the tropical Atlantic ocean. Limnol. Oceanogr. 11:614–620.

Gomont, M. 1892. Monographie des Oscillariees. Ann. Sci. Nat., Bot., Ser. 7, 15:263–368; 16:91–264; 16 pl.

Gomont, M. 1899. Sur quelques Oscillariees nouvelles. Bull. Sci. Bot. France 46:25–40.

Gorham, P. R. 1960. Toxic water blooms of blue-green algae. Can. Vet. J. 1:235.

Gorham, P. R. 1962. Laboratory studies on the toxins produced by water blooms of blue-green algae. Amer. J. Publ. Health 52:2100–2105.

Govindjee and B. Z. Braun. 1974. Light absorption, emission, and photosynthesis. *In* W. D. P. Stewart (Editor), Algal Physiology and Biochemistry. Botanical Monographs Vol. 10, pp. 346–390. Cambridge.

Granhall, U. 1976. The presence of cellulose in heterocyst envelopes of blue-green algae and its role in relation to nitrogen fixation. Physiol. Plant 38:208–216.

Gray, B. E., C. A. Lipschultz, and E. Gantt. 1973. Phycobilisomes from a blue-green alga, *Nostoc* sp. J. Bact. 116:471–478.

Grunow, A. 1867. Algae. Reise der Osterreichischen Fregatte "Norvara" um die Erde in den Jahren 1857-1858-1859. Bot. Theil, 1:1–104. Vienna.

Hagedorn, H. 1961. Untersuchungen uber die Feinstruktur der Blaualgen. Z. Naturforsch. 16b:825–29.

Halfen, L. N. and R. W. Castenholz. 1971. Gliding motility in the blue-green alga *Oscillatoria princeps*. J. Phycol. 7:133–145.

Hamm, D. and H. J. Humm. 1976. Benthic algae of the Anclote estuary. II. Bottom-dwelling species. Florida Sci. 39:209–29.

Harvey, W. H. 1858. Nereis Boreali Americana. Part III. Chlorospermae. Smithsonian Contributions to Knowledge 10(1):1–140, pl. 35–50. Washington, D.C.

Hauck, F. 1878. Beitrage zur Kenntnis der adriatischen Algen. Osterr. Bot. Zeitschr. 8:77–81.

Hauck, F. 1885. Die Meeresalgen Deutschlands und Osterreichs, *In* L. Rabenhorst, Kryptogamen-Flora von Deutschlands, Osterreich und der Schweiz. Second Edition, Vol. 2. Leipzig. 575 pp., 236 figs., 5 pl.

Holt, S. C., J. M. Shively, and J. W. Greenawalt. 1974. Fine structure of selected species of the genus *Thiobacillus* as revealed by chemical fixation and freeze etching. Can. J. Microbiol. 20:1347–1351.

Howe, M. A. 1914. The marine algae of Peru. Mem. Torrey Bot. Club 15:1–185.

Howe, M. A. 1920. Algae. *In* N. L. Britton and C. F. Millspaugh (Editors), The Bahama Flora.

Howe, M. A. 1924. Notes on algae of Bermuda and the Bahamas. Bull. Torrey Bot. Club 51: 351–359.

Hoyt, W. D. 1917-1918. Marine algae of Beaufort, N. C., and adjacent regions. Bull. Bureau of Fisheries (U.S.) 36:367–556.

Hulburt, E. M., J. H. Ryther, and R. L. Guillard. 1960. The phytoplankton of the Sargasso Sea off Bermuda. J. Conseil Perm. Internat. Explor. Mer 25:115–28.

Humm, H. J. 1952. Notes on the marine algae of Florida. I. The intertidal rocks at Marineland. Fla. State Univ. Studies 7:17–23.

Humm, H. J. 1963. Some new records and range extensions of Florida marine algae. Bull. Mar. Sci. 13:516–526.

Humm, H. J. 1964. Epiphytes of the sea grass *Thalassia testudinum* in Florida. Bull. Mar. Sci. 14:306–341.

Humm, H. J. 1976. The benthic algae of Biscayne Bay. Proc. Biscayne Bay Symposium I, pp. 71–93. Univ. of Miami Sea Grant Special Report No. 5.

Humm, H. J. 1979. The Marine Algae of Virginia. University Press of Virginia, Charlottesville. 263 pp., 103 figs.

Humm, H. J. and R. L. Caylor. 1957. The summer marine flora of Mississippi Sound. Publ. Inst. Mar. Sci., Univ. Texas 4:228–264.

Humm, H. J. and R. M. Darnell. 1959. A collection of marine algae from the Chandeleur Islands. Publ. Inst. Mar. Sci., Univ. Texas 6:265–276.

Humm, H. J. and H. H. Hildebrand. 1962. Marine algae from the Gulf coast of Texas and Mexico. Publ. Inst. Mar. Sci., Univ. Texas 8:227–268.

Humm, H. J. and C. R. Jackson. 1955. A collection of marine algae from Guantanamo Bay, Cuba. Bull. Mar. Sci. 5:240–246.

Ingram, L. O. 1973. Occurrence of cell lytic enzymes in blue-green bacteria. J. Bact. 116:832–35.

Ingram, L. O., D. Pierson, J. F. Kane, C. van Baalen, and R. A. Jensen. 1972. Documentation of auxotrophic mutation in blue-green bacteria: Characterization of tryptophan auxotroph in *Agmenellum quadruplicatum*. J. Bact. 111:112–118.

Ingram, L. O. and W. D. Fisher. 1972. Selective inhibition of DNA synthesis by 2-deoxyadenosine in the blue-green bacterium *Agmenellum quadruplicatum*. J. Bact. 112:170–175.

Ingram, L. O. and W. D. Fisher. 1973. Stimulation of cell division by croton oil in blue-green bacteria. J. Bact. 114:874–875.

Ingram, L. O., C. van Baalen, and J. A. Calder. 1973. Role of reduced exogenous organic compounds in the physiology of the blue-green bacteria (algae): Photoheterotrophic growth of an "autotrophic" blue-green bacterium. J. Bact. 114:701–705.

Jarosch, R. 1962. Gliding. *In* R. A. Lewin (Editor), Physiology and Biochemistry of Algae, Chapter 36. Academic Press, New York.

Jensen, T. E. 1968. Electron microscopy of pholphosphate bodies in a blue-green alga, *Nostoc pruniforme*. Arch. Mikrobiol. 62:144–152.

Jensen, T. E. 1969. Fine structure of developing polyphosphate bodies in a blue-green alga, *Plectonema boryanum*. Arch. Mikrobiol. 67:328–338.

Jensen, T. E. and L. M. Sicko. 1974. Phosphate metabolism in blue-green algae. I. Fine structure of the "polyphosphate overplus" phenomenon in *Plectonema* boryanum. Can. J. Microbiol. 20:1235–1239.

Joly, A. B. 1957. Contribucao ao conhecimento da flora ficologica marinha de Baia de Santos e Arredores. Boletim 17, Botanica 14, Fac. de Philos., Cien. et Lett., Univ. de Sao Paulo, Brasil. 199 pp., 19 pl.

Jones, D. D. and M. Jost. 1970. Isolation and chemical characterization of gas vacuole membranes from *Microcystis aeruginosa*. Arch. Mikrobiol. 70:43–64.

Jost, M. 1965. Die Ultrastruktur von *Oscillatoria rubescens*. Arch. Mikrobiol. 50:211–45.

Jurgens, G. H. B. 1816–1822. Algae aquaticae quas in littore maris Dynastium Jeveranam et Frisiam orientalem alluentis rejectas et in harum terrarum aquis habitantes collegit. Decades 1–20. Jever.

Kapp, R., S. E. Stevens, Jr., and J. L. Fox. 1975. A survey of available nitrogen sources for the growth of the blue-green alga *Agmenellum quadruplicatum*. Arch. Mikrobiol. 104:135–138.

Kessel, M. 1977. Identification of a phosphorus-containing storage granule in the cyanobacterium *Plectonema boryanum* by electron microscope X-ray microanalysis. J. Bact. 129:1502–1505.

Klebahn, H. 1895. Gasvakuolen, ein Bestandteil der Zellen der Wasserbluthebildenden Phycochromaceen. Flora Oder Allgem. Bot. Zeit. 80:241–1282.

Kumar, H. D. 1962. Apparent genetic recombination in a blue-green alga. Nature 196:1121–1122.

Kumar, H. D. 1964. Adaptation of a blue-green alga to sodium selenate and chloramphenicol. Plant Cell Physiol. 5:465–472.

Kützing, F. T. 1843. Phycologia generalis, oder Anatomie, Physiologie und Systemkunde der Tange. Leipzig. xxxiii + 458 p., 80 pl.

Kützing, F. T. 1845. Phycologica germanica d. i. Deutschlands Algen in bundigen Beschreibungen, nebst einer Anleitung zum Untersuchen und Bistimmen dieser Gewachse fur Anfanger. Nordhausen. x + 240 pp.

Kützing, F. T. 1849. Species Algarum. Leipzig. vi + 922 pp.

Kützing, F. T. 1849–69. Tabulae Phycologicae. Vol. I-19. Nordhausen. 1200 pl.

Kützing, F. T. 1863. Diagnosen und Bemerkungen zu drei und siebenzig neuen Algenspecies.

Lagerheim, G. 1883. Bidrag till Sveriges Algflora. Ofvers. Kgl. Svensk.-Akad. Forhandl 40(2):37–78.

Lagerheim, G. 1886. Note sur le *Mastigocoleus*, noveau genre des algues marines de l'ordre des Phycochromacees. Notarisia 1:65–69.

Lang, N. J. 1965. Electron microscopic study of heterocyst development in *Anabaena azollae*. J. Phycol. 1:127–134.

Lang, N. J. 1968. The fine structure of blue-green algae. Ann. Rev. Microbiol. 22:15–46.

Lang, N. J. 1976. Speculations on a possible essential function of the gelatinous sheath of blue-green algae. Can. J. Microbiol. 22:1181–1185.

Lang, N. J. and P. Fay. 1971. The heterocysts of blue-green algae. II. Details of ultrastructure. Proc. Roy. Soc. London, Ser. B, 178:193–203.

Lang, N. J. and P. M. M. Rae. 1967. Structures in a blue-green alga resembling prolamellar bodies. Protoplasma 64:67–74.

Lang, N. J. and B. A. Whitton. 1973. Arrangement and structure of thylakoids. *In* N. G. Carr and B. A. Whitton (Editors), The Biology of Blue-Green Algae. Bot. Monograph 9, pp. 66–79. Blackwell, London.

Leak, L. V. 1967. Fine structure of the mucilaginous sheath of *Anabaena* sp. J. Ultrastructure Res. 21:61–74.

Lemmermann, E. 1905. Die Algenflora der Sandwich-Inseln. Bot. Jahrb. Syst., Pflanzen-geschichte und Pflanzengeog. 34:607–663.

Lewin, R. A. 1974. Biochemical taxonomy. *In* W. D. P. Stewart (Editor), Algal Physiology and Biochemistry. Bot. Monographs Vol. 10, pp. 29–30. Univ. Calif. Press, Berkeley.

Lewin, R. A. 1977. Prochloron, type genus of the Prochlorophyta. Phycologia 16:217.

Lewin, R. A. and L. Cheng. 1975. Associations of microscopic algae with didemnid ascid-ians. Phycologia 14:149–152.

Lindstedt, A. 1943. Die Flora marinen Cyanophycees der schwedischen Westkuste. Lund. 121 pp., 11 pl.

Lyngbye, H. C. 1819. Tentamen hydrophytologiae Danicae. Copenhagen. 248 pp., 70 pl.

Meeuse, B. J. D. 1962. Storage products. *In* R. A. Lewin (Editor), Physiology and Biochemis-try of Algae, Chapter 18, p. 289–303. Academic Press, New York.

Meyen, F. J. F. 1839. Neues System der Pflanzenphysiologie, Vol. 3. Berlin. 627 pp.

Mueller, T. M. 1976. The distribution and seasonality of marine algae of coastal Louisiana and the adjacent offshore continental shelf. Master's thesis, Dept. of Marine Science, Univ. of South Florida, 151 pp.

Nadal, N. G. M. 1971. Sterols of *Spirulina maxima*. Phytochemistry 10:2537–2538.

Nageli, C. 1849. Gattungen einzelliger Algen physiologisch und systematisch bearbeitet. Neue Denckschriften der allgemeinen schweitzerischen Gesellschaft fur die gesammten Naturwissenschaften, Vol. 10. Zurich.

Newton, Lily. 1931. A Handbook of British Seaweeds. British Museum, London. xiii + 478 pp., 270 figs.

Nielsen, C. S. 1954. The multitrichomate Oscillatoriaceae of Florida. Quart. J. Florida Acad. Sci. 17:25–42.

Oersted, A. S. 1842. Beretning om en Excursion til Trindelen, alluvial Dannelse i Odensef-jord. Nat. Tidskrift (København) 17:552–68.

Padan, E. and M. Shilo. 1973. Cyanophages—Viruses attacking blue-green algae. Bact. Rev. 37:343–370.

Pankratz, H. S. and C. C. Bowen. 1963. Cytology of blue-green algae. I. The cells of *Symploca muscorum*. Amer. J. Bot. 50:387–399.

Pelroy, R. A., M. R. Kirk, and J. A. Bassham. 1976. Photosystem II: regulation of macro-molecule synthesis in the blue-green algae *Aphanocapsa* 6714. J. Bact. 128:623–632.

Perkins, R. D. and I. Tsentas. 1976. Microbiol infestation of carbonate substrates planted on the St. Croix shelf, West Indies. Bull. Geol. Soc. Amer. 87:1615–1628.

Prescott, G. W. 1968. The Algae: A Review. Houghton Mifflin, New York, 436 pp.

Purdy, E. G. and L. S. Kornicker. 1958. Algal disintegration of Bahamian limestone coasts. J. Geol. 66:96–99.

Rabenhorst, L. 1865. Flora Europea Algarum. Second edition. Leipzig. 319 pp.

Randall, J. E. 1958. A review of ciguatera, tropical fish poisoning, with a tentative explana-tion of its cause. Bull. Mar. Sci. 8:236–267.

Rayss, T. 1959. Contribution a la connaissance de la flore marine de la mer Rouge. Bull. 23, Sea Fisheries Res. Sta., Haifa, Israel. 32 pp.

Reinsch, P. F. 1875. Contributiones algologiam et fungologiam. Leipzig. 104 pp., 88 pl.

Reitz, R. C., and J. G. Hamilton. 1968. The isolation and identification of two sterols from two species of blue-green algae. Comp. Biochem. Physiol. 25:401–416.

Rosenvinge, L. 1893. Grønlands Havalger. Medd. om Grønland, Vol. 3. Copenhagen.

Safferman, R. S. and M. E. Morris. 1963. Algal virus: Isolation. Science 140:679–80.

Safferman, R. S. and M. E. Morris. 1967. Observations on the occurrence, distribution, and seasonal incidence of blue-green algal viruses. Appl. Microbiol. 15:1219–1222.

Salton, M. R. J. 1964. The Bacterial Cell Wall. Elsevier, New York. 293 pp., 23 figs.

Sauvageau, C. 1895. Sur le Radaisa, nouveau genre de Myxophycée. J. de Bot. 9:7.

Schmidt, J. 1901. Plankton fra det Rode Hav og Adenbugten. Vidensk. Meddel. fra d. Naturh. Foren. Copenhagen.

Schopf, J. W. 1970. Electron microscopy of organically preserved precambrian fossils. J. Paleontol. 44:1–6. Seoane-Camba, J. 1965. Estudios sobre las algas bentonicas en la costa sur de la Peninsula Iberica. Invest. Pesquera 29:3–216.

Setchell, W. A. 1895. Notes on some Cyanophyceae of New England. Bull. Torrey Bot. Club 22:427.

Setchell, W. A. and N. L. Gardner. 1903. Algae of northwestern America. Univ. Calif. Publ. Bot. 1:165–418.

Shear, H. and A. E. Walsby. 1975. An investigation into the possible light-shielding role of gas vacuoles in a planktonic blue-green alga. Brit. Phycol. J. 10:241–251.

Shestakov, S. F. and N. T. Khyen. 1970. Evidence for genetic transformation in the blue-green alga *Anacystis nidulans*. Mol. Genetics 107:372–375.

Shilo, M. 1966. Predatory bacteria. Science J. (London) 2:33–77.

Shilo, M. 1970. Lysis of blue-green algae by *Myxobacter*. J. Bact. 104:453–461.

Shimura, S., and Y. Fujita. 1975. Phycoerythrin and photosynthesis of the pelagic blue-green alga *Trichodesmium thiebautii* in the waters of Kuroshio, Japan. Marine Biol. 31:121–128.

Shively, J. M., F. L. Ball, and B. W. Kline. 1973. Electron microscopy of the carboxysomes of *Thiobacillus neapolitanus*. J. Bact. 116:1405–1411.

Sicko-Goad, L. and T. E. Jensen. 1976. Phosphate metabolism in blue-green algae: II. Changes in phosphate distribution during starvation and the polyphosphate overplus phenomenon in *Plectonema boryanum*. Amer. J. Bot. 63:183–188.

Simon, R. D. 1971. Cyanophycin granules from the blue-green alga *Anabaena cylindrica*: a reserve material consisting of co-polymers of aspartic acid and arginine. Proc. Nat. Acad. Sci. 68:265–267.

Simon, R. D. 1973. Measurement of the cyanophycin granule polypeptide contained in the blue-green alga *Anabaena cylindrica*. J. Bact. 116:1213–1216.

Singh, R. N. and R. Sinha. 1965. Genetic recombination in the blue-green alga, *Cylindrospermum majus*. Nature 207:782–783.

Stanier, R. Y., R. Kunisawa, M. Mandel, and G. Cohen-Bazire. 1971. Purification and properties of unicellular blue-green algae. Bact. Rev. 35:171–205.

Stewart, J. R. and R. M. Brown. 1969. *Cytophaga* that kills and lyses algae. Science 165: 1523–1524.

Stewart, W. D. P. and G. A. Codd. 1975. Polyhedral bodies (carboxysomes) of nitrogen-fixing blue-green algae. Brit. Phycol. J. 10:273–278.

Stewart, W. D. P., G. P. Fitzgerald, and R. H. Burris. 1968. Acetylene reduction by nitrogen-fixing blue-green algae. Arch. Mikrobiol. 62:336–348.

Stewart, W. D. P. and M. Lex. 1970. Nitrogenase activity in the blue-green alga *Plectonema boryanum* strain 594. Arch. Mikrobiol. 73:250–260.

Stewart, W. D. P. and H. N. Singh. 1975. Transfer of nitrogen-fixing (NIF) genes in the blue-green alga *Nostoc muscorum*. Biochem. Biophys. Res. Commun. 62:62–69.

Strodtman, S. 1895. Die Ursache der Schwebevermögens bei den Cyanophyceen. Biol. Zentralbl. 15:113–115.

Tanaka, Y., H. Matsuguchi, and T. Katayama. 1974. Comparative biochemistry of carotinoids in algae: IV. Carotinoids in Cyanophyta, blue-green algae, *Spirulina platensis*. Mem, Fac. Fish., Kagoshima Univ. 23:111–116.

Taylor, B. F., C. C. Lee, and J. S. Bunt. 1973. Nitrogen fixation associated with the marine blue-green alga *Trichodesmium*, as measured by the acetylene-reduction technique. Arch. Mikrobiol. 88:205.

Taylor, W. R., 1928. Marine algae of Florida with special reference to the Dry Tortugas. Publ. Carnegie Inst. of Wash. No. 379; Papers from the Tortugas Lab. 25:1–219, 3 figs., 8 tables, 37 pl.

Thuret, G. 1875. Essai de classification des Nostochinees. Ann. Sci. Nat., Bot., ser. 6, 1:372–382.

Tilden, Josephine E. 1910. Minnesota Algae, Vol. I. Minneapolis. iv + 319 pp., 20 pl.

Tyagi, V. V. 1975. The heterocysts of blue-green algae. Biol. Rev. 50:247–284.

Umezaki, I. 1961. The marine blue-green algae of Japan. Mem, Coll. Agr. Kyoto Univ. 83, Fish., ser. 8, 149 p.

Utkilin, H. C. 1976. Thiosulfate as an electron donor in the blue-green alga, *Anacystis nidulans*. J. Gen. Microbiol. 95:177–180.

Van Baalen, C., and R. M. Brown, Jr. 1969. The ultrastructure of the marine blue-green alga *Trichodesmium erythraeum* with special reference to the cell wall, gas vacuoles, and cylindrical bodies. Arch. Mikrobiol. 69:79–91.

Van Gorkom, H. J., and M. Donze. 1971. Localization of nitrogen fixation in *Anabaena*. Nature 234:231–232.

Vickers, Anna. 1905. Liste des algues marines de la Barbade. Ann. Sci. Nat., Bot., ser. 9, 1:45–66.

Walsby, A. E. 1972. Structure and function of gas vacuoles. Bact. Rev. 36:1–32.

Weller, D., W. Doemel, and T. D. Brock. 1975. Requirements of low oxidation-reduction potential for photosynthesis in a blue-green alga (*Phormidium* sp.). Arch. Mikrobiol. 104:7–14.

West, W. 1899. Some Oscillatoriaceae from the plankton. J. Bot. (London) 37:337–338.

White, A. W., and M. Shilo. 1975. Heterotrophic growth of the filamentous blue-green alga *Plectonema boryanum*. Arch. Mikrobiol. 102:123–127.

Wildman, R. B., and C. C. Bowen. 1974. Phycobilisomes in blue-green algae. J. Bact. 117:866–881.

Williams, L. G. 1948. Seasonal alternation of marine floras at Cape Lookout, N.C. Amer. J. Bot. 35:682–695.

Winkenbach, F., and C. P. Wolk. 1973. Activities of enzymes of the oxidative and the reductive pentose phosphate pathways in heterocysts of a blue-green alga. Plant Physiol. 52:480–483.

Wolk, C. P., and P. W. Shaffer. 1976. Heterotrophic micro- and macrocultures of a nitrogen-fixing cyanobacterium. Arch. Mikrobiol. 110:145–148.

Wyatt, J. T., and J. K. G. Silvey. 1969. Nitrogen fixation by *Gloeocapsa*. Science 165:908–909.

Zanardini, G. 1858. Plantarum in Mari Rubro hucusque collectarum enumeratio. Mem. dell'Imperiale Reale Inst. Veneto di Sci., Lett. Arti 7:209–309, pl. 3–14.

Zanardini, G. 1862–1876. Scelta di Ficee Nuove o Piu Rare del Mari Adriatico. Mem. dell'Imperiale Reale Inst. Veneto di Sci., Lett. Arti. Vol. 9–19.

TAXONOMIC INDEX

acervatus, Xenococcus, 119
acuminata, Oscillatoria, 128
aeruginea, Calothrix, 151
aeruginea, Fremyella, 154
aeruginea, Microchaete, 155
aerugineum, Merismopedium, 53
aeruginosa, Anacystis, 57
aestuarii, Lyngbya, 132
Agmenellum, 52
amethystea, Pleurocapsa, 115
amoena, Oscillatoria, 128
Amphithrix, 146
Anabaena, 162
Anabaina, 88, 161
Anacystis, 55
Aphanocapsa, 110
Aphanothece, 109
aponina, Gomphosphaeria, 53
arenaria, Schizothrix, 73
Arthrospira, 74
atlantica, Symploca, 139
atra, Rivularia, 159
autumnale, Phormidium, 137

balani, Brachytrichia, 93
battersii, Plectonema, 142
biscayensis, Dermocarpa, 113
bonnemaisonii, Oscillatoria, 125
borgesenii, Hydrocoleum, 79
bornetiana, Dichothrix, 157
bornetiana, Rivularia, 159
Brachytrichia, 92
brevis, Arthrospira, 76
brevis, Oscillatoria, 127

caespitosa, Hyella, 118
calcicola, Schizothrix, 71
calothrichoides, Plectonema, 142
Calothrix, 84
cantharidosmum, Hydrocoleum, 144

cantharidosmus, Hydrocoleus, 144
chaetomorphae, Xenococcus, 118
Chamaesiphonaceae, 58
Chlorogloea, 61
Chondrocystis, 60
Chroococcaceae, 47
Chroococcus, 105
chthonoplastes, Microcoleus, 145
coadunata, Rivularia, 159
Coccochloris, 48
Coccogonales, 45
commune, Nostoc, 164, 165
comoides, Hydrocoleum, 143
conferta, Entophysalis, 61
confervicola, Calothrix, 150
confervoides, Lyngbya, 133
conglomerata, Gloeocapsa, 56
consociata, Calothrix, 149
contarenii, Calothrix, 150
contortum, Trichodesmium, 124
convoluta, Merismopedia, 53
convolutum, Merismopedium, 53
corallinae, Oscillatoria, 126
crepidinum, Gloecapsa, 110
crepidinum, Pleurocapsa, 115
crosbyanum, Phormidium, 135
crustacea, Calothrix, 84, 151
Cyanobacteria, 4
Cyanophyta, 44

Dermocarpa, 111
deusta, Entophysalis, 60
Dichothrix, 155
dimidiata, Anacystis, 57

elabens, Coccochloris, 50
elabens, Microcystis, 51
elegans, Merismopedia, 109
endophytica, Chlorogloea, 61

189

endophytica, Entophysalis, 61
enteromorphoiodes, Hormothamnion, 161
Entophysalis, 59, 116
entophytum, Nostoc, 165
epiphytica, Lyngbya, 131
erythraea, Oscillatoria, 67
erythraeum, Trichodesmium, 124

fasciculata, Calothrix, 152
fragile, Phormidium, 136
Fremyella, 153
fucicola, Dichothrix, 156
fuliginosa, Pleurocapsa, 115
fuliginosum, Scytonema, 160
fusca, Nodularia, 51
fuscolutea, Gloeocapsa, 56
fusco-violacea, Calothrix, 149

glauca, Merismopedia, 108
glaucum, Merismopedium, 53
Gloeocapsa, 110
Gloeothece, 49
glutinosum, Hydrocoleum, 144
Gomphosphaeria, 53
gracilis, Lyngbya, 131
granulosa, Entophysalis, 117
grisea, Fremyella, 154
grisea, Microchaete, 154

harveyana, Nodularia, 167
hawaiiensis, Nodularia, 167
hofmannii, Scytonema, 85
Hormactis, 93
Hormogonales, 62
Hormothamnion, 89, 160
hydnoides, Calothrix, 140
hydnoides, Symploca, 139
Hydrocoleum, 142
Hyella, 118
Hypheothrix, 146

inaequalis, Anabaena, 163
intracellularis, Richelia, 153

janthina, Amphithrix, 147
Johannesbaptistia, 51

kurzii, Porphyrosiphon, 77
kurzii, Sirocoleum, 79

lacustris, Schizothrix, 146
laetevirens, Oscillatoria, 128
laete-viridis, Symploca, 139
leibleiniae, Dermocarpa, 114
licheniformis, Anabaina, 88
linckia, Nostoc, 165
linearis, Gloeothece, 49
longiarticulata, Hypheothrix, 146
longiarticulata, Schizothrix, 146
longifila, Calothrix, 154
longifila, Fremyella, 154
lutea, Chlorogloea, 61
lutea, Lyngbya, 69
lutea, Oscillatoria, 68
Lyngbya, 129
lyngbyacea, Oscillatoria, 83
lyngbyaceum, Hydrocoleum, 143
lyngbyaceus, Microcoleus, 81

magnoliae, Entophysalis, 117
major, Spirulina, 122
majuscula, Lyngbya, 133
margaritifera, Oscillatoria, 126
marina, Anacystis, 56
marina, Aphanocapsa, 56, 110
marina, Oncobyrsa, 111
Mastigocoleus, 91
meneghiniana, Lyngbya, 131
meneghiniana, Spirulina, 122
Merismopedia, 106
Merismopedium, 53
mexicana, Schizothrix, 73
Microchaete, 153
Microcoleus, 80
Microcrocis, 47
Microcystis, 51
microscopicum, Nostoc, 164
miniata, Arthrospira, 77
miniata, Oscillatoria, 126
miniata, Spirulina, 77
miniatus, Porphyrosiphon, 77
minima, Dermocarpa, 114
minutus, Chroococcus, 106
montana, Anacystis, 56
Myxophyceae, 3

neapolitana, Arthrospira, 74
neapolitana, Oscillatoria, 75

nigro-viridis, Oscillatoria, 127
nitida, Rivularia, 158
Nodularia, 166
nordstedtii, Spirulina, 122
Nostoc, 89, 164
Nostocaceae, 83
nostocorum, Plectonema, 141
notarisii, Porphyrosiphon, 79
novum, Merismopedium, 53

ocellatum, Scytonema, 160
olivacea, Dermocarpa, 114
olivaceus, Sphaenosiphon, 114
Oncobyrsa, 111
oscillarioides, Anabaina, 88
oscillatoria, 66, 124
Oscillatoriaceae, 64

pallida, Aphanothece, 109
parasitica, Calothrix, 149
parietina, Calothrix, 84
pellucida, Johannesbaptistia, 51
penicillata, Dichothrix, 156
penicillatum, Phormidium, 137
peniocystis, Gloeocapsa, 49
persicinum, Phormidium, 136
Phormidium, 134
pilosa, Calothrix, 152
Plectonema, 140
Pleurocapsa, 114
polyotis, Rivularia, 158
Porphyrosiphon, 76
prasina, Dermocarpa, 113
Procaryotae, 4
profunda, Symploca, 138
prolifera, Calothrix, 151
Protophyta, 4
pulvinata, Calothrix, 150
punctata, Merismopedia, 109

quadruplicatum, Agmenellum, 53
quoyi, Nostoc, 93

Raphidiopsis, 83
rhizosolenia, Calothrix, 153
Richelia, 153
Rivularia, 157
Rivulariaceae, 62, 63

rivulariarum, Lyngbya, 131
rosea, Dermocarpa, 113
rupestris, Gloeothece, 49
rupicola, Dichothrix, 157

sabulicola, Merismopedia, 108
salina, Aphanocapsa, 56
salinarum, Oscillatoria, 128
schauinslandii, Chondrocystis, 60
Schizothrix, 70, 145
schousboei, Xenococcus, 119
scopulorum, Calothrix, 151
Scytonema, 85, 160
Scytonemaceae, 62
semiplena, Lyngbya, 132
Sirocoleum, 79
Skujaella, 123
smaragdina, Dermocarpa, 114
smaragdinus, Sphaenosiphon, 114
solitaria, Dermocarpa, 113
solutum, Hormothamnion, 161
sordida, Lyngbya, 132
Sphaenosiphon, 114
sphaericum, Nostoc, 164
Spirulina, 65, 119
spongeliae, Phormidium, 137
spumigena, Nodularia, 90, 167
spumigena, Nostoc, 89
stagnina, Aphanothece, 49, 110
stagnina, Coccochloris, 49
Stigonemataceae, 62, 90
submembranacea, Oscillatoria, 69
submembranaceum, Phormidium, 138
subsalsa, Spirulina, 66, 123
subtilissima, Spirulina, 121
subuliforme, Phormidium, 137
subuliformis, Oscillatoria, 127
Symploca, 138

tenerrima, Spirulina, 121
tenerrimus, Microcoleus, 73, 145
terebrans, Plectonema, 141
testarum, Mastigocoleus, 91
thermale, Agmenellum, 53
thermalis, Merismopedia, 53
thiebautii, Trichodesmium, 124
tolypothrichoides, Scytonema, 160
torulosa, Anabaena, 163

Trichodesmium, 123
turgidus, Chroococcus, 57, 106

vaginatus, Microcoleus, 81
variabilis, Anabaina, 163
versicolor, Spirulina, 122
violacea, Amphithrix, 147
violacea, Dermocarpa, 112

violacea, Entophysalis, 117
vitiensis, Microchaete, 153
vivipara. Calothrix, 152

warmingiana, Merismopedia, 109
williamsii, Oscillatoria, 77

Xenococcus, 118

SUBJECT INDEX

Acetylene Reduction Test, 17
Akinete, 26
Anthocyanins, 23
ATP, 20, 25

Bacteria, 4, 6, 33
Bergey's Manual, 4
Bibliography, 178
Black zone, 34

Carboxysomes, 19, 27
Carotinoids, 11
Cell wall, 21
Chemoautotrophism, 19
Chlorophyll, 12
Chromoplasm, 8
Ciguatera, 39
Classification, 4, 62
Collecting advice, 39
Cyanophage, 33
Cyanophycin granules, 16, 27
Cytology, 8

Dissemination, 25
Distribution, 34
DNA, 8, 21
Drouet, Dr. Francis, 2, 3, 5

Endospores, 10, 25
Euphotic Zone, 35

Fibrils, 21, 24
Food sources, 13
Function:
   of gas vacuoles, 30
   of heterocysts, 29
   of sheath, 24

Gas vacuoles, 29-32

Genetic recombination, 14
Genome, 8, 10
Glossary, 174
Glycoaminopeptides, 21
Glycogen, 16

Heterocysts, 19, 27
Heterotrophic growth, 13
Hormogonia, 25, 26
Hot springs, 38

Intertidal Zone, 34

Keys to Genera, 44, 102

Limestone, 36
   boring into, 36, 60, 71, 85, 87, 91
   precipitation of, 37
Lipids, 13, 15, 17
Lysis of bacteria, 22
Lysozyme, 21, 29

Microscopic examination, 40
Morphology, 6
Motility, 24
Mucopolysaccharides, 23
Muramic acid, 21
Mutation, 13
Myxophycean starch, 13, 16

Nitrogen fixation, 17, 19, 67
Nitrogen metabolism, 15
Nitrogenase, 17, 19, 20
Nucleoplasm, 8

Organic nitrogen sources, 15

Photosynthesis, 12
Photosystem II, 12, 19, 20, 27
Phycobilins, 10, 27

Pigments, 10-12, 18, 23, 27
Plankton, 36
Planococci, 26
Plasmodesmata, 27, 29
Polar bodies, 29
Polyhedral bodies, 19, 27
Polyphosphate granules, 16, 27
Polysaccharides, 23
Preservation, 39

Red tide, 36, 67
Reproduction, 25
Ribosomes, 9, 27
RNA, 8

Salinity, 37
Sea saw dust, 36, 67
Sheath, 22-24
Sterols, 17
Stored food, 13, 15-17

Taxonomy, 5
Temperature, 37
Thylakoids, 10, 27
Toxic bluegreens, 38

Virion, 34

Xanthophyll, 23